# Selected Topics on Polynomials

# Selected Topics on Polynomials

Andrzej Schinzel
of the Mathematics Institute of
the Polish Academy of Sciences

Ann Arbor
The University of Michigan Press

ISBN 978-0-472-08026-7 (sqb)
Library of Congress Catalog Card No. 81-71930
Published in the United States of America by
The University of Michigan Press and simultaneously
in Rexdale, Canada, by John Wiley & Sons Canada, Limited
Printed and bound by CPI Group (UK) Ltd, Croydon, CR0 4YY

Paperback ISBN: 978-0-472-75194-5

# Preface

The present book originated in a course of lectures given at the University of Michigan in Winter Term, 1977. The notes of these lectures have been taken by Dr. Patrick Morton. Professor Donald J. Lewis proposed publication of the notes in an expanded form by the University of Michigan Press and devoted much of his time to correcting and polishing of my manuscript. Three anonymous referees have made valuable critical comments. The typing has been done by Ms. E. Banulewicz and Ms. E. Hitczenko from the Mathematics Institute of the Polish Academy of Sciences and by Ms. A. Velez from the department of mathematics of the University of California, Irvine. My warm thanks go to all the above persons for their help and to the institutions for their financial support during the preparation of the book.

# Contents

# Introduction

In these lectures I have gathered some results on polynomials scattered in the literature that recommended themselves to me by the simplicity of formulation notwithstanding a certain depth. I have added several results of my own that seemed to me to have similar features. The book is divided into two parts: algebraic and arithmetic. Polynomials considered in the first part have coefficients in an arbitrary field, sometimes algebraically closed, those considered in the second part in an algebraic number field. In the first part the only knowledge assumed from the beginning is that of volume 1 of van der Waerden's *Algebra* (English edition), in the second part the rudiments of algebraic number theory as exposed e.g. in H.Mann's book *Introduction to algebraic number theory* are assumed. I also make use of elementary calculus and two results from the theory of analytic functions: Parseval's identity and Puiseux's theorem. (In the proof of Theorem 28 a theorem about quadratic forms is used in order to simplify the calculations, however the result of these calculations can be verified without the said theorem. In counterexamples but not in the proofs some special results about Diophantine equations are used). The order of the discussed topics is logical in the sense that theorems proved earlier are often used in the proofs of results given later but not vice versa. Here is the list of topics in Part I.

1. Lüroth's theorem and its consequences for polynomials in many variables (§§ 2,3) .
2. Ritt's theory of the composition of polynomials in one

variable (§§ 4,5) with application to diophantine equations (§ 6).

3. Kronecker's theorem on the connection between coefficients of factors and of the product of polynomials (§ 7).
4. Reducibility of polynomials typified by $F(x,y) + G(z)$ (§ 8).
5. Elimination theory for systems of homogeneous equations with applications to the algebra of polynomials (§§ 9, 10).
6. Polynomials reducible in an algebraically closed field for every value of some of the variables (§ 11).
7. Equivalence between reducibility of a polynomial in one variable over a field and solvability of a suitable equation in the same field (§ 12).
8. Capelli's criterion for the reducibility of binomials, its applications and an extension (§§ 13, 14, 17).
9. Reducibility of polynomials of the form $F(x_1^{n_1},\dots,x_k^{n_k})$ (§§ 15, 16).

Let me say something more about these topics.

1. Lüroth's theorem is well known and included with proof in van der Waerden's book. I give a different proof that is longer but constructive and I present the consequences Lüroth's theorem has for fields of rational functions in many variables, of transcendence degree 1. These were discovered by Gordan 1887 and rediscovered by Igusa 1951. The much more difficult problem of the minimal number of generators for fields of rational functions of transcendence degree greater than 1 belongs properly to algebraic geometry and here only references are given.

2. Ritt 1923 gave a complete analysis of the behaviour of polynomials in one variable with complex coefficients under

composition (superposition). He called a polynomial $F$ prime if it is not the composition of two polynomials of lower degree and proved the two main results.

(i) In every representation of a polynomial as the composition of prime polynomials the number of factors is the same and their degrees coincide up to a permutation.

(ii) If $A,H$ and $B,G$ are polynomials of relatively prime degrees $m$ and $n$ respectively and

$$(*) \qquad A(G) = B(H)$$

then $A,B,G,H$ can be given explicitly.

Ritt showed also how every representation of $F$ as the composition of prime factors can be obtained from a given one by solving several equations of the form (*), where $A$ and $B$ are prime.

Many years later Engstrom 1941 simplified Ritt's proof of (i) and extended this result to polynomials over an arbitrary field of characteristic 0. Levi 1942 did the same for the result (ii) assuming however that $A,B,G,H$ are prime and that the field in question is algebraically closed, arguing moreover that the latter assumption causes no loss of generality. Both Engstrom and Levi used Lüroth's theorem in their proofs. Interest in the subject was renewed by Fried and Mac Rae 1969 who rediscovered Engstrom's results and extended them to polynomials $F$ over fields of characteristic $p$ of degree not divisible by $p$. A counterexample showing that without the latter restriction (i) fails has been given by Dorey and Whaples 1974. The latter authors rediscovered Levi's results simplifying the proofs considerably. They also assumed in

(ii) that A,B,G,H are prime. This assumption has not been made by Fried 1974, however his assertion about A,B,G,H is less precise than Ritt's. In the present notes § 4 contains a proof of Fried and Mac Rae's form of (i) on Engstrom's lines and § 5 contains a proof of (ii) for polynomials over an arbitrary field of characteristic 0 or of characteristic $p > \max\{m,n\}$. In particular the field is not assumed to be algebraically closed, which matters as far as the explicit form of the solutions of (*) is concerned. Ritt's term "prime" is replaced throughout by "indecomposable".

The application of these results to Diophantine equations given in § 6 is based on the celebrated Siegel theoren characterizing curves with infinitely many integer points: this theorem is not proved in the text and the result of § 6 (as well as those of §§ 4, 5) is not used in the sequel.

3. The theorem of Kronecker, once called fundamental and now nearly forgotten is proved in § 7 for the sake of applications in §§ 8 and 10. Its name refers to the fact that it could be used as a basis for establishing the arithmetic of ideals in algebraic fields.

4. H.Davenport and the author proved in 1963 that a polynomial of the form $F(x,y) + G(z)$ is reducible in a field $k$ of characteristic 0 iff $F = H(A(x,y))$ $A,H \in k[t]$ and $H(t) + G(z)$ is reducible in $k$. § 8 contains a natural generalization of this result and a discussion of related results of Tverberg. The difficult problem of reducibility of $H(t) + G(z)$ is not pursued, however references are given. The results of § 8 are not used in the sequel.

5. Elimination theory for systems of homogeneous equations is presented here not for the sake of itself but to prepare tools to be used in §§ 11 and 24. The shortest available proofs have been chosen for the existence of resultant systems and of the resultant.

6. Polynomials $F(t,x,y)$ with complex coefficients that are linear in $t$ and irreducible, but become reducible as polynomials in two variables for every value of $t$ are characterized by Bertini's theorem. This theorem about linear systems of reducible varieties has been extensively studied in algebraic geometry. Here we adopt an algebraic point of view and in § 11 prove an extension of Bertini's theorem to polynomials of an arbitrary degree in $t$ following Krull's 1937 paper. The results of § 11 are not used in the sequel.

7. The equivalence between reducibility of polynomials and solvability of equations is proved here both for its intrinsic interest and for intended application in the proof of Hilbert's irreducibility theorem in § 22. It is for this purpose that it was originally established by Mertens 1911.

8. Capelli 1898 gave a simple necessary and sufficient condition for reducibility of a binomial $x^n - a$ over a field of characteristic 0. The case of positive characteristic was settled by Rédei 1959. Capelli's original proof is given in § 13 with the slight modification needed to handle the case of positive characteristic. The theorem can also be viewed as a necessary and sufficient condition for an element a of a field k to satisfy the

equality $[k(\sqrt[n]{a}):k]=n$. In this aspect the theorem is open to generalisation, namely one can study the degree $[k(\sqrt[n_1]{a_1}, \sqrt[n_2]{a_2},\ldots,\sqrt[n_\ell]{a_\ell}):k]$. An all encompossing result in this direction for separable extensions has been found by M.Kneser 1974. It is reproduced in § 17 together with a more immediate extension of Capelli's theorem.

The proof of Kneser's theorem does not use Capelli's theorem; therefore of least for fields of characteristic 0 one can extract from § 17 a proof of the latter theorem independent from that of § 13.

It is an almost immediate consequence of Capelli's theorem that for $a \neq 0$ the polynomial $x^m + y^n + a$ is irreducible in every field of characteristic 0 containing a. This observation is generalized in § 14 to an easily applicable irreducibility criterion for polynomials in many variables.

9. Following the work of Ritt 1931, Gourin 1933 proved that for a given polynomial $F(x_1,\ldots,x_s)$ with more than two terms, irreducible over the complex field, and for arbitrary positive integers $t_1,\ldots,t_s$ the factorisation of $F(x_1^{t_1},\ldots,x_s^{t_s})$ into irreducible factors can be derived from the factorisation of $F(x_1^{\tau_1},\ldots,x_s^{\tau_s})$, where $\langle\tau_1,\ldots,\tau_s\rangle$ belongs to a finite set of integral vectors depending only on F. Gourin's proof applies with small modifications to polynomials over an arbitrary algebraically closed field and integers $t_1,\ldots,t_s$ not divisible by the characteristic of the field. In this slightly modified form it is reproduced in § 15.

An extension of the theorem to polynomials over fields

no longer algebraically closed is given in § 16. It turns out that the condition on the polynomial $F$ to have more than two terms has to be replaced by the condition that $F$ is not of the form

$$(**) \qquad F_o(x_1^{\delta_1} x_2^{\delta_2} \cdots x_s^{\delta_s}) \prod_{r=1}^{s} x_r^{-d \min\{0,\delta_r\}}$$

where $F_o(x)$ is a polynomial of degree $d$ and $\delta_1,\ldots,\delta_s$ are integers possibly negative. How to deal with polynomials of type (**) in the case of algebraic number fields is shown in § 18 belonging already to Part II of the book.

Besides this topic in Part II the following subjects are expounded.

10. Estimates for the product of zeros of a polynomial outside the unit circle with application to problems of reducibility (§§ 19, 20, 21).

11. Hilbert's irreducibility theorem and related problems about Diophantine equations with parameters (§§ 22-28).

Here is a brief introduction to these quite independent subjects. According to a classical theorem of Kronecker if a monic irreducible polynomial $F(x) \neq x$ with integral coefficients has no zero outside the unit circle then it must be cyclotomic. It has been asked by Lehmer 1933 whether for non-cyclotomic polynomials $F$ in question the product $M(F)$ of the relevant zeros can be arbitrarily close to 1. This problem remains unsolved till now, but Smyth 1971 has made great progress towards the solution by proving that for non-reciprocal polynomials $F$ in question $M(F) \geq \theta_o$, where $\theta_0$ is the real zero of $x^3 - x - 1$. An exposition of Smyth's work is given in § 19 and an extension to polynomials over totally real or some totally complex fields in § 20. In § 21 these results are combined with the known upper bounds for $M(F)$ in order to estimate the number of irreducible non-reciprocal factors of a polynomial in terms of the sum of squares of its coefficients.

Hilbert's irreducibility theorem in nearly its simplest from asserts that if a polynomial $F(x,t)$ is irreducible over a number field $K$ as polynomial in two variables then the set $T$ of integers $t^*$ such that $F(x,t^*)$ is irreducible in $K$ is infinite. § 22 contains a proof of Hilbert's original, much more general form of this theorem, by a method due to Dörge. The proof involves the Puiseux theorem from the theory of analytic functions and Mertens's theorem from § 12. It turns out incidentally that the set T contains almost all integers in the sense of density. In § 23 another property of the set $T$ is proved again in a more general context, namely that it contains a whole arithmetic progression. Via Mertens's theorem there is a close conection between Hilbert's theorem and the following statement.

If an equation $F(x,t^*) = 0$ is solvable for $x$ in $K$ for all but finitely many integers $t^*$ then there exists a rational function $x(t) \in K(t)$ satisfying the equation $F(x(t),t) = 0$.

It is a natural question to ask whether a similar theorem holds for equations with more than one parameter or more than one unknown. This question, not entirely solved yet, is discussed in §§ 23-25. It turns out for instance that if

$f \in K(t)$ and the equation $f(t^*) = x^2 + y^2$ is solvable for $x, y \in K$ for all integers $t^*$ then $f(t) = x(t)^2 + y(t)^2$, where $x(t), y(t) \in K(t)$.

This is the starting point for § 26, where a similar result is established with $x^2 + y^2$ replaced by a norm of an extension of prime degree of a number field. However the form $x^2 + y^2$ has also the property that if it represents over $K$ the value $F(t^*)$ of a polynomial $F \in K[t]$ for all integers $t^*$ then

$$F(t) = X(t)^2 + Y(t)^2,$$

where $X, Y \in K[t]$.
Other norm forms with a similar property are found in § 27 by means of a theorem of Bauer 1916, which permits one to characterize normal extensions $K/k$ by the set of prime ideals of the ground field $k$ that have in $K$ a prime ideal factor of relative degree 1. In § 28 the topic is brought to a close by exhibiting a class of extensions called Bauerian, which behave in the above respect similarly to normal extensions.

The theorems and the important definitions are in the whole book numbered successively irrespective of sections, on the other hand ad hoc definitions (called conventions) and formulae are numbered separately for each section. Lemmata and Corollaries are subordinate to definitions and theorems. References to the bibliography placed at the end of the book are given by author's name and the year of publication, in rare cases, where there are two papers of the same author in one year they are distinguished by letters a,b.

# Notation

In Part I the letter $k$ is reserved for fields, in Part II where only algebraic number fields are considered they are denoted by $K$.

$k^*$ is the multiplicative group of the field $k$.

$\hat{k}$ is the algebraic closure of $k$.

$\mathbb{Q}, \mathbb{R}, \mathbb{C}$ are the fields of rational, real and complex numbers, respectively.

$\mathbb{F}_q$ is the finite field of $q$ elements.

$\mathbb{Z}$ is the ring of rational integers.

$\mathbb{N}$ is the set of positive integers.

$I_\ell$ is the identity matrix of size $\ell$.

$\zeta_m$ is a primitive $m^{th}$ root of unity.

Bold face letters denote vectors. If distinct bold face letters occur as arguments of a polynomial it is assumed that the components of the relevant vectors are independent variables. For a polynomial $F(\mathbf{x}_1, \mathbf{x}_2, \ldots, \mathbf{x}_n)$ over an integral domain $D$:

$|F|_{\mathbf{x}_i}$ is the maximum degree of $F$ with respect to all variables occuring in $\mathbf{x}_i$. If $n = 1$, $|F|_{\mathbf{x}_i} = |F|$;

$\deg_{\mathbf{x}_i} F$ is the degree of $F$ viewed as a polynomial in $\mathbf{x}_i$, if $n = 1$, $\deg_{\mathbf{x}_1} F = \deg F$;

If $g = \frac{F}{G}$, where $F, G$ polynomials such that $(F, G) = 1$ then

$$\deg g = \max(\deg F, \deg G).$$

$\operatorname{disc}_x F$ is the discriminant of $F$ with respect to the variable $x$; if this is the only variable $\operatorname{disc}_x F = \operatorname{disc} F$;

$$F(x_1,\dots,x_n) \underset{D}{\overset{\text{can}}{=\!=}} \text{const} \prod_{\rho=1}^{s} F_\rho(x_1,\dots,x_n)^{e_\rho}$$

means that polynomials $F_\rho$ are irreducible in $D$ and pairwise relatively prime.

For a polynomial $F$ over $\mathbb{C}$

$\|F\|$ is the sum of the squares of the absolute values of the coefficients of $F$.

$C(F)$ is the content of $F$, defined as the highest common divisor of the coefficients of $F$ if they algebraic.

For polynomials $F$ and $G$ in one variable we set

$$F \circ G = F(G(x)).$$

For a polynomial $F \in \mathbb{C}[x]$, $F(x) = a_0 \prod_{i=1}^{n} (x - \alpha_i)$, we set

$$M(F) = |a_0| \prod_{i=1}^{n} \max(1, |\alpha_i|).$$

For a rational function of the form

$$f(x_1,x_2,\dots,x_n) = x_1^{\alpha_1} x_2^{\alpha_2} \dots x_n^{\alpha_n} F(x_1,\dots,x_n)$$

where $F$ is a polynomial prime to $x_1x_2\dots x_n$ we set

$$Jf(x_1,x_2,\dots,x_n) = F(x_1,x_2,\dots,x_n).$$

Parenthesis $(.,.)$ denotes the highest common divisor, $k(\ )$ denotes a field extension of the field $k$. Brackets $[.,.]$ denote the lowest common multiple, $D[\ ]$ denotes a polynomial ring extension of the ring $D$.

Brackets <...> denote vectors, G< > denotes an extension of the group G.

Bar denotes the complex conjugation; dash usually, but not always differentiation.

For a finite field extension $K/k$ the symbol $N_{K/k}$ denotes the norm from $K$ to $k$ or from $K(x)$ to $k(x)$, where $x$ is transcendental over $k$.

$\mathrm{ord}_p a$ is the highest power to which a prime element $p$ of a unique factorization domain or a prime ideal $p$ of a Dedekind domain divides an element $a$ of this domain.

## Section 1. Auxiliary results on transcendental and inseparable extensions

Here we have gathered some notions and theorems contained essentially in van der Waerden 1970 but not given there explicitly.

Definition 1. *Elements* $\xi_1, \ldots, \xi_m$ *of a field* $K$ *algebraically independent over a subfield* $k$ *of* $K$ *form a transcendence basis of* $K/k$ *if* $K/k(\xi_1, \ldots, \xi_n)$ *is algebraic.*

Theorem A. (Steinitz 1910). *If* $\{\xi_1, \ldots, \xi_n\}$ *is a transcendence basis of* $K/k$ *and* $\eta_1, \ldots, \eta_m \in K$ *are algebraically independent over* $k$, *then* $\xi$*'s can be renumbered in such a way that* $\{\eta_1, \ldots, \eta_m, \xi_{m+1}, \ldots \xi_n\}$ *is a transcendence basis for* $K/k$.

Corollary 1. If $K/k$ has a finite transcendence basis then all transcendence bases of $K/k$ have the same number of elements called the transcendence degree of $K/k$ (tr.deg. $K/k$).

Corollary 2. For arbitrary $\xi_1, \ldots, \xi_n$ we have

$$\text{tr.deg } k(\xi_1, \ldots, \xi_n) \leq n.$$

Theorem B. (Steinitz 1910). *If* $k \subset K \subset L$ *and* tr.deg $L/k$ *is finite then*

$$\text{tr.deg } L/k = \text{tr.deg } K/k + \text{tr.deg } L/K.$$

Proofs. see van der Waerden 1970, § 10.4.

Theorem C. (Steinitz 1910). *Let* $K$ *be an extension of a field* $k$. *Elements of* $K$ *separable over* $k$ *form a field called the separable closure of* $k$ *in* $K$.

Definition 2. *The separable closure of* $k$ *is the separable closure of* $k$ *in its algebraic closure* $\hat{k}$.

Definition 3. *A monic polynomial* $F \in k[x]$ *irreducible in* $k$ *is purely inseparable over* $k$ *if it is of the form* $x^{p^e} - a$; $p = \text{char } k$. *An element algebraic over* $k$ *is purely inseparable over* $k$ *is it is a zero of a purely inseparable polynomial over* $k$. *An extension of* $k$ *is purely inseparable if all its elements are purely inseparable over* $k$.

Theorem D. (Steinitz 1910). *Every algebraic extension* $K$ *of a field* $k$ *is purely inseparable over the separable closure of* $k$ *in* $K$.

Proofs. see van der Waerden 1970, § 6.8.

## Section 2. Lüroth's theorem

We first prove

Theorem 1. (E.Noether 1915 for char $k=0$, Samuel 1952 in general). *If* $k \subset K \subset k(x_1,\ldots,x_n)$, *then* $K = k(g_1,\ldots,g_t)$, *where the* $g_i$ *lie in* $k(x_1,\ldots,x_n)$. *If char* $k = 0$, $t \leq 1 + tr.deg.K/k$.

Proof. By Theorem B $r = \text{tr deg } K/k \leq n$. Let $\{g_1,\ldots,g_r\}$ be a transcendence basis of $K/k$. By Theorem A, one can renumber the $x$'s so that $\{g_1,\ldots,g_r, x_{r+1},\ldots,x_n\}$ is a transcendence basis for $k(x_1,\ldots,x_n)/k$. We assert that $[K:k(g_1,\ldots,g_r)] \leq [K(x_{r+1},\ldots,x_n) : k(g_1,\ldots,g_r,x_{r+1},\ldots,x_n)] < \infty$. Suppose that inequality not true, so we have $y_1,\ldots,y_s \in K$, linearly independent over $k(g_1,\ldots,g_n)$, but linearly dependent over $k(g_1,\ldots,g_r,x_{r+1},\ldots,x_n)$; thus

$$b_1y_1 + \ldots + b_sy_s = 0,$$

where

$$b_i = \sum_{j\in N^{n-r}} a_{ij}x_{r+1}^{j_{r+1}}\cdots x_n^{j_n}, \quad a_{ij} \in k(g_1,\ldots,g_r)$$

We can write this as $\sum_{j\in N^{n-r}} x_{r+1}^{j_{r+1}}\cdots x_n^{j_n} \sum_{i=1}^{s} a_{ij}y_i = 0$, whence $\sum_{i=1}^{s} a_{ij}y_j = 0$.

By the assumption $a_{ij} = 0$, for all $i,j$ so $b_i = 0$ for all $i \leq s$. Thus our assertion is proved and we take $g_{r+1},\ldots,g_t$ to be generators of $K$ over $k(g_1,\ldots,g_r)$. If char $K = 0$, we need to add only one generator by Abel's theorem of primitive element.

From Cor.1 to Theorem A we know that $t \geq tr.deg.K/k$.

Lüroth's theorem states that in the case $n = 1$, we have here an equality.

Theorem 2. (Lüroth 1876 for $k = \mathbb{C}$, Steinitz 1910 in general). *If* $k \subset K \subset k(x)$ *and* $K \neq k$ *then* $k = k(y)$, $y$ *a rational function of* $x$.

Proof (Netto 1895). By Theorem 1 we have $K = k(g_1, \ldots, g_s)$, $g_i \in k(x)$ $k$. Let $g_i = \frac{F_i}{G_i}$, where $(F_i, G_i) = 1$, $\deg F_i > \deg G_i$. Consider the polynomials

$$F_i(t) - g_i G_i(t) \in k(x)[t], \qquad (i = 1, \ldots, s)$$

and let their highest common factor be $\frac{D(x,t)}{d_o(x)}$, where $D(x,y)$ is primitive as a polynomial in $t$ with the leading coefficient $d_o(x)$. By Gauss's Lemma, $D(x,t) \mid F_i(t)G_i(x) - F_i(x)G_i(t)$, so $F_i(t)G_i(x) - F_i(x)G_i(t) = D(x,t)C_i(x,t)$. Take $i$ such that $\deg g_i = m$ is least. If $D(x,t)$ is of lower degree in $t$ (i.e. $< m$) than is $F_i(t) - g_i G_i(t)$, then $\deg_t C_i > 0$. Suppose $\deg_x C_i(x,t) = 0$, say $C_i(x,t) = C_i(t)$. Let $F_i(t) \equiv \tilde{F}_i(t) \pmod{C_i(t)}$, $\deg \tilde{F}_i < \deg C_i$, similarly $G_i(t) \equiv \tilde{G}_i(t) \pmod{G_i}$, $\deg \tilde{G}_i < \deg C_i$. We have $\tilde{F}_i(t)G_i(x) - F_i(x)\tilde{G}_i(t) \equiv 0 \pmod{C_i(t)}$ and comparing degrees in $t$ we get $\tilde{F}_i(t)G_i(x) = F_i(x)\tilde{G}_i(t)$. But $(F_i, G_i) = 1$, hence either $F_i \in k$ or $\tilde{F}_i(t) = 0$ and either $G_i \in k$ or $\tilde{G}_i(t) = 0$. All four resulting cases are impossible, since $\deg g_i > 0$ and $(F_i, G_i) \not\equiv 0 \pmod{C_i}$. Hence $C_i$ depends on both $x, t$, and $\deg_x D < m$. Now $\frac{D(x,t)}{d_o(x)}$ is monic in $k(x)[t]$. Its coefficients belong to $K$, have degree $< m$ and at least one coefficient must be non--constant since $D \notin k[t]$. Add one of the non-constant coefficients to the generators $g_1, \ldots, g_s$ and repeat the whole procedure.

By repeating the procedure with the larger set of generators, we must come to a point, where

$$(1) \qquad \underset{i \le s}{\text{g.c.d.}}[F_i(t) - g_i(x)G_i(t)] = F_\nu(t) - g_\nu(x)G_\nu(t).$$

Then $g_\nu(x)$ is the required generator. Indeed, for each $i$

$$F_i(t) - g_i(x)G_i(t) = (F_\nu(t) - g_\nu(x)G(t))C_i(t), \quad C_i \in k(x)[t] .$$

Now in $k(g_\nu)[t]$ for a given $i$ there exist $P,Q,R,S$ such that

$$F_i(t) = P(t)[F_\nu(t) - g_\nu G_\nu(t)] + Q(t), \ \deg Q < \deg[F_\nu(t) - g_\nu G_\nu(t)]$$

$$G_i(t) = R(t)[F_\nu(t) - g_\nu G_\nu(t)] + S(t), \quad \deg S < \deg[F_\nu(t) - g_\nu G_\nu(t)].$$

If $Q = 0$, $F_i(t) = P(t)[F_\nu(t) - g_\nu G_\nu(t)]$ and writing $P(t)$ as $\dfrac{T(g_\nu,t)}{p(g_\nu)}$, where $T,p$ polynomials over $k$ we get

$$F_i(t) = \frac{T(g_\nu,t)}{p(g_\nu)}[F_\nu(t) - g_\nu G_\nu(t)],$$

$$F_i(t)p(g_\nu) = T(g_\nu,t)[F_\nu(t) - g_\nu G_\nu(t)],$$

which is impossible, since $F_\nu(t) - g_\nu G_\nu(t)$ does not factor in $k(g_\nu,t)$.
Hence $Q \neq 0$ and similarly $S \neq 0$. Also

$$F_i(t) - g_i G_i(t) = [P(t) - g_i R(t)][F_\nu(t) - g_\nu G_\nu(t)] + Q(t) - g_i S(t).$$

If follows from (1) that $Q(t) = g_i S(t)$. Taking the leading coefficients $q_o$, $s_o$ of $Q, S$ respectively we get $q_o = g_i s_o \in k(g_\nu)$, so $g_i = \frac{q_o}{s_o} \in k(g_\nu)$ Q.E.D.

The above proof is constructive, that is permits one to find the generator of $K$ given as $k(g_1, \ldots, g_s)$ and express $g_1, \ldots, g_s$ in terms of this generator. A short non-constructive proof can be found in v.d.Waerden 1970. If tr.deg.$K/k = 2$ and $k = \mathbb{C}$ is algebraically closed then in analogy with Lüroth's theorem $K = k(g_1, g_2)$ for suitable $g_1, g_2$ (Castelnuovo 1894), however if $k = \mathbf{Q}$ or $\mathbb{R}$ this need not be the case (Segre 1951). The case of algebraically closed fields of positive characteristic is treated by Zariski 1958. If tr.deg.$K/k = 3$ then even for $k = \mathbb{C}$ $K/k$ may need four generators. (Artin and Mumford 1971, Clemens and Griffiths 1971 and Iskovich and Manin 1971). For an extension of Lüroth's theorem in a different direction see Moh and Heinzer 1979.

Theorem 3. (Gordan 1887 for char $k = 0$, Igusa 1951 in general). *If* $k \subset K \subset k(x_1,\ldots,x_n)$, *tr deg* $K/k = 1$, *then* $K = k(g)$.

Proof. (Samuel 1952, valid for $k$ infinite). By Theorem 1, $K = k(\varphi_1,\ldots,\varphi_t)$.

By Theorem A, on renumbering $x$'s one can assume $x_2,\ldots,x_n$ are algebraically independent over $K$. We have

$$k(x_2,\ldots,x_n) \subset K(x_2,\ldots,x_n) \subset k(x_1,\ldots,x_n).$$

By Lüroth's theorem

$$K(x_2,\ldots,x_n) = k(x_2,\ldots,x_n,\eta).$$

Hence

$$\varphi_i = g_i(\eta,x_2,\ldots,x_n), \quad \text{where} \quad g_i \in k(X_1,\ldots,X_n) \quad (1 \le i \le t)$$

and

$$\eta = h(\varphi_1,\ldots,\varphi_t,x_2,\ldots,x_n), \quad \text{where} \quad h \in k(y_1,\ldots,y_t,x_2,\ldots,x_n).$$

Therefore

$$\varphi_i = g_i(h(\varphi_1,\ldots,\varphi_t,x_2,\ldots,x_n),x_2,\ldots,x_n) \quad (1 \le i \le t)$$

identically over $K$ since $x_2,\ldots,x_n$ are algebraically independent over $K$. Choose values $x_2^*,\ldots,x_n^*$ in $k$ so that after substitution $x_i = x_i^*$ the rational function on the right hand side makes sense. Now

$$h(\varphi_1,\ldots,\varphi_t,x_2^*,\ldots,x_n^*)$$

is the desired generator for $K/k$, since

$$\varphi_i = g_i(h(\varphi_1,\ldots,\varphi_t,x_2^*,\ldots,x_n^*),\ x_1^*,\ldots,x_n^*) \text{ for all } i \leq t.$$

Q.E.D.

Theorem 4. (E.Noether 1915 for char $k = 0$, Schinzel 1963 in general). *If under assumption of Theorem 3 $K$ contains a nonconstant polynomial over $k$ then $K$ has a generator which is a polynomial $\in k[x_1,\ldots,x_n]$.*

We recall that for a polynomial $F$ in one variable, $|F|$ is the degree of $F$.

Lemma. Let $(R(x), S(x)) = 1$, $R(x,y) = y^{|R|}R(\frac{x}{y})$, $S(x,y) = y^{|S|}S(\frac{x}{y})$, $R(0)\,S(0) \neq 0$ and $(P(x), Q(x)) = 1$. Then

$Q$, $R(P,Q)$, $S(P,Q)$ are prime in pairs.

Proof. We write $RU + SV = 1$, where $U,V$ are polynomials in one variable of degrees less than $|S|$ and $|R|$ respectively and get

$$R(x,y)U(x,y) + S(x,y)V(x,y) = y^{|RU|}.$$

Now we substitute $x = P, y = Q$ and get

$$R(P,Q)U(P,Q) + S(P,Q)V(P,Q) = Q^{|RU|}.$$

Lemma follows since $(R(P,Q),Q) = (Q,S(P,Q)) = 1$.

Proof of Theorem 4. Let $g = \frac{P(\mathbf{x})}{Q(\mathbf{x})}$, $(P,Q) = 1$.

By hypothesis there is a polynomial $F$ in $K$,

$$F = \frac{R}{S}(g) = \frac{R(P/Q)}{S(P/Q)} = \frac{R(P,Q)Q^{-r}}{S(P,Q)Q^{-s}}Q^{s-r}.$$

By the lemma, $S(P,Q) \in k$, $Q \in k$ or $s \geq r$. Factoring $S(P,Q)$ we get

$$S(P,Q) = \alpha(P - \xi_1 Q)(P - \xi_2 Q)\ldots(P - \xi_s Q) \in k^*, \quad \text{hence}$$

$$P - \xi_i Q = \gamma_i \in \hat{k}.$$

Now $S(x)$ cannot have two different roots, since otherwise $(\xi_1 - \xi_2)Q = -\gamma_1 + \gamma_2$ implies successively $Q \in \hat{k}$, $P \in \hat{k}$, $g \in \hat{k}$, which is impossible. Thus

$$S(P,Q) = \alpha(P - \xi Q)^s, \quad P - \xi Q = \gamma.$$

If $Q \notin k$ then $\xi \in k$, $\gamma \in k$,

$g = \frac{P}{Q} = \xi + \frac{\gamma}{Q}$ is expressed as a rational function of $Q$ and we take $Q$ to be a generator.

If $Q \in k$, we may take $P$ to be a generator.

## Section 4. Ritt's first theorem

Convention. Capital letters denote polynomials in one variable.

Theorem 5. (Engstrom 1941 for char k = 0, Fried and MacRae 1969 in general). *If* $k(F) \cap k(G)$ *contains a polynomial* $H$ *such that* $|H| \not\equiv 0 \bmod \operatorname{char} k$, *then*

$$[k(F) : k(F) \cap k(G)] = \frac{[|F|, |G|]}{|F|},$$

$$[k(F, G) : k(F)] = \frac{|F|}{(|F|, |G|)}.$$

Lemma 1. If $H \in k(F)$, then $H = A(F)$.

Proof. If $H = \frac{R}{S}(F)$, $(R,S) = 1$ then $(R(F), S(F)) = 1$ and hence $S$ = constant.

Lemma 2. $[k(x) : k(F)] = |F|$.

Proof. If $F = a_0x^n + \ldots + a_n$, then $G = a_0X^n + \ldots + a_n - F$ is an irreducible polynomial over $k(F)$ with $x$ as a zero.

Lemma 3. If $A \in k[x]$ is monic and $|A| = rn$, $r \not\equiv 0(\operatorname{char} k)$, then there exists a monic polynomial $C \in k[x]$ such that $|C| = n$, $|A - C^r| < n(r-1)$.

Proof. For each nonnegative $i \leq n$ there exists a $C_i \in k[x]$ such that $|C_i| = n$, $|A - C_i^r| < nr-i$. We prove this by induction on $i$. If $i = 0$, $C_0 = x^n$. Suppose proved for $i-1$, where $0 < i \leq n$. Hence we have a polynomial $C_{i-1}$ of degree $n$ such that $|A - C_{i-1}^r| < nr - i+1$. Look for $C_i$ of form

$$C_i = C_{i-1} + \xi x^{n-i}.$$

We have

$$C_i^r = C_{i-1}^r + r\, C_{i-1}^{r-1}\xi x^{n-i} + \binom{r}{2} C_{i-1}^{r-2}\, \xi^2 x^{2(n-i)} + \ldots$$

The degree of the 3-rd and latter terms is at most $n(r-2) + 2(n-i) = nr-2i < nr-i$. Consider $A - C_{i-1}^r - r\xi C_{i-1}^{r-1}x^{n-i}$. We have $|C_{i-1}^{r-1}x^{n-i}| = nr-i$. Select $\xi$ so that terms of degree $nr-i$ cancel each other and then $|A - C_i^r| < nr-i$. Since $C_o$ is monic the construction insures all $C_i$ are monic.

Proof of Theorem 5. By Lemma 1 and the hypothesis we have $H = A(F) = B(G)$ by Lemma 1. Without loss of generality we may assume $H, F, G, A, B$ all monic. If $|F| = n = d\nu$; $|G| = m = d\mu$, where $(\nu,\mu) = 1$ we have $|H| = r\, d\mu\nu$, $r \not\equiv 0(\text{char } k)$.

Also, $|A| = r\mu$, $|B| = r\nu$.

By Theorem 4, there exists a polynomial generating $k(F) \cap k(G)$; by Lemma 1 we may assume it to be $H$. We shall prove $r = 1$. By Lemma 3 there exist monic polynomials $C, D \in k[x]$ such that $|C| = \mu$, $|D| = \nu$, $|A-C^r| < \mu(r-1)$, $|B-D^r| < \nu(r-1)$.

Hence

$$|A(F) - C^r(F)| < d\mu\nu(r-1),$$

$$|B(G) - D^r(G)| < d\mu\nu(r-1),$$

$$|C^r(F) - D^r(G)| < d\mu\nu(r-1).$$

But $(C(F) - D(G))(C^{r-1}(F) + \ldots + D^{r-1}(G)) = C^r(F) - D^r(G)$. Since $C, D, F, G$ are monic and $r \not\equiv 0$ (mod char $k$), the second factor has actual degree $(r-1)d\mu\nu$ and therefore $C(F) = D(G)$ and clearly they lie in $k(F) \cap k(G) = k(H)$. Then $|C(F)| \geq |H|$, i.e. $d\mu\nu \geq rd\mu\nu$, whence $r = 1$.

Hence $[k(F) : k(F) \cap k(G)] = [k(F) : k(H)] = [k(F) : k(A(F))] =$

$= |A| = \mu = \frac{[|F|,|G|]}{|F|}$. Similarly $[k(G) : k(F) \cap k(G)] =$

$= \frac{[|F|,|G|]}{|G|}$ and since the right hand sides are coprime

$$[k(F,G) : k(F) \cap k(G)] = \frac{[|F|,|G|]}{(|F|,|G|)}.$$

The theorem follows.

The following examples shows that the assumption $|H| \not\equiv 0 \bmod \text{char } k$ cannot be omitted.

Example 1. $k = \mathbb{F}_2$, $F = x^2$, $G = x^2 + x$, $H = x^4 + x^2 = F^2 + F = G^2$; $k(F) \cap k(G) = k(H)$, $k(F,G) = k(x)$.

Example 2 (Bremner and Morton 1978). $k = \mathbb{F}_3$, $F = x^2$, $G = x^2 + x$, $H = x^6 + x^4 + x^2 = F^3 + F^2 + F = G^3 + G^2$; $k(F) \cap k(G) = k(H)$, $k(F,G) = k(x)$.

Corollary 1. $[k(F) : k(F) \cap k(G)] = [k(F,G) : k(G)]$, if $|H| \not\equiv 0 \bmod \text{char } k$.

The second example given above shows that the assumption $|H| \not\equiv 0$ (mod char k) cannot be omitted here either.

Definition 4. *A polynomial* $F$ *is indecomposable over* $K$ *iff* $F = F_1 \circ F_2$, $F_1, F_2 \in k[x]$ *implies* $|F_1|$ *or* $|F_2| = 1$.

Corollary 2. (Fried and MacRae 1969). If $|F| \not\equiv 0$ (mod char k) and $F$ is indecomposable over $k$, then it is indecomposable over any extension of $k$.

Proof. Let $F = F_1 \circ F_2$ be a decomposition of $F$ over some extension $K$ of $k$, $|F_1| = r$, $|F_2| = n$.
Assume $F_1$ is monic. If $F_1 = x^r + a_1 x^{r-1} + \ldots + a_r$ we can write $F = \tilde{F}_1 \circ \tilde{F}_2$ where $\tilde{F}_1(x) = F_1(x - \frac{a_1}{r})$, $\tilde{F}_2 = F_2(x) + \frac{a_1}{r}$ and the coefficient of $x^{r-1}$ in $\tilde{F}_1(x)$ is 0.
By Lemma 3 there exists a $C \in k[x]$ such that $|C| = n$ and $|F - C^r| < n(r-1)$, so $|\tilde{F}_1 \circ \tilde{F}_2 - C^r| < n(r-1)$. It follows

that $|\tilde{F}_2^r - C^r| < n(r-1)$. However $\tilde{F}_2^r - C^r = \prod_{\nu=1}^{r} (\tilde{F}_2 - \zeta_r^\nu C)$ and at most one factor has degree $< n$.

It follows that $\tilde{F}_2 = \zeta_r^\nu C$, where $\zeta_r^\nu \tilde{F}_2 \in k[x]$ for some $\nu \leq r$.

Setting $\tilde{F}_1(x) = x^r + \sum_{i=1}^{r} b_i x^{r-i}$ we infer from $F = \tilde{F}_1 \circ \tilde{F}_2$ by induction on i that $b_i \zeta_r^{-\nu i} \in k$, whence $\tilde{F}_1(\zeta_r^\nu x) \in k[x]$. But then $F$ is decomposable over $k$.

Example. Let $k = \mathbb{F}_2$. Then $F(x) = x^4 + x^2 + x = (x^2 + \alpha x)^2 + \alpha^{-1}(x^2 + \alpha x)$, where $\alpha^3 - \alpha + 1 = 0$, $\alpha \in \mathbb{F}_8$ shows that the assumption $|F| \not\equiv 0 \pmod{\text{char } k}$ cannot be omitted.

Corollary 3. $F \in k[x]$ is indecomposable over $k$ if and only if the extension $k(x)/k(F)$ is primitive i.e. if and only if $k(F) \subset K \subset k(x)$ implies $K = k(F)$ or $k(x)$.

Proof. Suppose $k(F) \subset K \subset k(x)$. Then by Theorem 4, $K = k(G)$ and hence by Lemma 1 to Theorem 5 $F = H(G)$. Thus $F$ is decomposed unless $|H| = 1$ or $|G| = 1$.

Definition 5. *Two decompositions of* $F$, *say* $F = F_1 \circ F_2 \circ \ldots \circ F_r$ *and* $F = G_1 \circ G_2 \circ \ldots \circ G_r$ *are equivalent, symbolically* $\langle F_1, \ldots, F_r \rangle \sim \langle G_1, \ldots, G_r \rangle$ *or* $\langle F_i \rangle_{i \leq r} \sim \langle G_i \rangle_{i \leq r}$ *if there exist linear functions* $L_1, \ldots, L_{r-1}$, *such that* $G_1 = F_1 \circ L_1$, $G_j = L_{j-1}^{-1} \circ F_j \circ L_j$ $(1 < j < r)$, $G_r = L_{r-1}^{-1} \circ F_r$.

Theorem 6. (Ritt 1922 for $k = \mathbb{C}$, Engstrom 1941 for char k=0, Fried and MacRae 1969 in general). *If* $|F| \not\equiv 0 \pmod{\text{char } k}$, *and* $F = G_1 \circ G_2 \circ \ldots \circ G_r = H_1 \circ H_2 \circ \ldots \circ H_s$, *where* $G_i, H_i$ *are indecomposable of degree* $> 1$, *then* $r = s$, *and the sequences* $\langle |G_i| \rangle, \langle |H_i| \rangle$ *are permutations of each other. Moreover, there exists a finite chain of decompositions* $F = F_1^{(j)} \circ \ldots \circ F_1^{(j)}$ $(j \leq n)$, *such that*

$$\langle F_i^{(1)}\rangle_{i\le r} = \langle G_i\rangle_{i\le r}, \quad \langle F_i^{(n)}\rangle_{i\le r} \sim \langle H_i\rangle_{i\le r}$$

*and*

(1) *for each* $j \le n$, $\langle F_i^{(j)}\rangle_{i\le r}$ *and* $\langle F_i^{(j+1)}\rangle_{i\le r}$ *differ only by having 2 consecutive terms with the same product (o) and distinct reversed degrees.*

Proof. (Induction on $|F|$). For $|F| = 1$ the theorem holds. Assume it is true for polynomials of degree $<|F|$ and let
$F = G_1 \circ G_2 \circ \dots \circ G_r = H_1 \circ H_2 \circ \dots \circ H_s$.
Case 1. $k(G_r) = k(H_s)$. Then $H_s = L \circ G_r$,
$G_1 \circ G_2 \circ \dots \circ G_{r-1} \circ G_r = H_1 \circ H_2 \circ \dots \circ H_{s-1} \circ L \circ G_r$.

If $A \circ B = C \circ B$, $|B| > 0$ then $A = C$. Thus
$G_1 \circ G_2 \circ \dots \circ G_{r-1} = H_1 \circ H_2 \circ \dots \circ (H_{s-1} \circ L)$ and by the inductive assumption $r - 1 = s - 1$; $r = s$. Moreover, there exists a chain of decompositions $\langle F_i^{(j)}\rangle_{i\le r-1}$ $(j \le n)$ satisfying (1) such that

$$\langle F_i^{(1)}\rangle_{i\le r-1} = \langle G_i\rangle_{i\le r-1}, \quad \langle F_i^{(n)}\rangle_{i\le r-1} \sim \langle H_i, \dots, H_{r-1} \circ L\rangle.$$

We set $F_r^{(j)} = G_r$ $(1 \le j \le n)$ and find
$\langle F_i^{(n)}\rangle_{i\le r} \sim \langle H_1, \dots, H_{r-2}, H_{r-1} \circ L, G_r\rangle \sim \langle H_i\rangle_{i\le r}$.

Case 2. $k(G_r) \ne k(H_s)$. Then $k(x) \supset k(G_r, H_s) \supsetneq k(G_r)$, thus by Cor.3 to Theorem 5 $k(G_r, H_s) = k(x)$.

By Cor.1 to Theorem 5 $[k(G_r) : k(G_r) \cap k(H_s)] =$
$[k(G_r, H_s) : k(H_s)] = [k(x) : k(H_s)] = |H_s|$.
Since $F \in k(G_r) \cap k(H_s)$, by Theorem 4 the intersection $k(G_r) \cap k(H_s)$ is generated by some polynomial $P$, hence

$P = A(G_r)$, $|A| = |H_s|$ and $P = B(H_s)$, $|B| = |G_r|$.

Suppose $A = A_1 \circ A_2$. Since $k(G_r) \cap k(H_s) = k(P)$,
$P \in k(A_2 \circ G) \cap k(H_s)$ implies $k(A_2 \circ G_r) \cap k(H_s) = k(P)$.
On the other hand $k(H_s) \subset k(H_s, A_2(G_r)) \subset k(x)$. Therefore,
either $k(H_s) \supset k(A_2 \circ G_r)$ or $k(H_s, A_2 \circ G_r) = k(x)$. In the
first case $k(P) = K(A_2(G_r))$ hence $|A_1| = 1$.
In the second case by Cor.1 to Theorem 5

$$[k(A_2(G_r)) : k(A_2(G_r)) \cap k(H_s)] = [k(x) : k(H_s)] = |H_s| = |A|,$$

$$[k(A_2(G_r)) : k(A_1(A_2(G_r)))] = |A|.$$

But the above degree also equals $|A_1|$; $|A_1| = |A|$ thus
$|A_2| = 1$.
It follows that $A, B$ are indecomposable.

We have now $F = C \circ P$ and if $C = C_1 \circ \ldots \circ C_t$,
$C_j$ indecomposable, it follows that

$$F = \begin{cases} C_1 \circ \ldots \circ C_t \circ A \circ G_r = G_1 \circ \ldots \circ G_{r-1} \circ G_r \\ \\ C_1 \circ \ldots \circ C_t \circ B \circ H_s = H_1 \circ \ldots \circ H_{s-1} \circ H_s, \end{cases}$$

hence $C_1 \circ \ldots \circ C_t \circ A = G_1 \circ \ldots \circ G_{r-1}$, $C_1 \circ \ldots \circ C_t \circ B =$
$H_1 \circ \ldots \circ H_{s-1}$ and by the inductive assumption $t + 1 = r - 1 =$
$s - 1$; $r = s$.
Moreover, there exists a chain of decompositions satisfying
(1) and such that

$$\langle F_i^{(1)} \rangle_{i \le r-1} = \langle G_i \rangle_{i \le r-1}$$

$$\langle F_i^{(n)} \rangle_{i \le r-1} \sim \langle C_1, \ldots, C_{r-2}, A \rangle.$$

It follows that for some linear function L

$$F_1^{(n)} \circ \ldots \circ F_{r-2}^{(n)} = C_1 \circ \ldots \circ C_{r-2} \circ L^{-1}, \quad F_{r-1}^{(n)} = L \circ A,$$

$$F_1^{(n)} \circ \ldots \circ F_{r-2}^{(n)} \circ (L \circ B) = C_1 \circ \ldots \circ C_{r-2} \circ B.$$

On the other hand, we have a chain of decompositions $\langle F_i^{(j)} \rangle_{i \leq r-1}$ $(n \quad j \quad n+m)$, whence

$$\langle F_1^{(n+1)}, \ldots, F_{r-1}^{(n+1)} \rangle = \langle F_1^{(n)}, \ldots, F_{r-2}^{(n)}, \quad L \circ B \rangle,$$

$$\langle F_1^{(n+m)}, \ldots, F_{r-1}^{(n+m)} \rangle \sim \langle H_1, \ldots, H_{s-1} \rangle.$$

$$\text{Define} \quad F_r^{(j)} = \begin{cases} G_r & \text{if} \quad j \leq n, \\ H_s & \text{if} \quad n < j \leq n+m. \end{cases}$$

The new chain satisfies all condition.

Without the assumption $|F| \not\equiv 0 \pmod{\text{char } k}$ Theorem 6 is not true in general, as it is shown by the following example due to Dorey and Whaples 1974:

$$F(x) = x^{p+1} \circ (x^p + x) \circ (x^p - x) = (x^{p^2} - x)^{p+1} =$$

$$= (x^{p^2} - x^{p^2-p+1} - x^p + x) \circ x^{p+1}.$$

If A,B,G,H are indecomposable, $G(A) = H(B)$ and $\langle G,A \rangle$ is not equivalent to $\langle H,B \rangle$ then by Theorem 5 $|G| = |B|$ and $|H| = |A|$ are relatively prime. Only the latter condition occurs in Ritt's second theorem, which deals with the above equation.

## Section 5. Ritt's second theorem

Definition 6. *Chebyshev's polynomials* $T_n(x)$ *are given by the recurrence formulae*

$$T_o(x) = 2,\ T_1(x) = x,\ T_{n+1}(x) = xT_n(x) - T_{n-1}(x).$$

Remark. The classical Chebyshev polynomials cos(n arc cos x) are equal to $\frac{1}{2}T_n(2x)$.

Corollary 1. $T_n(x)$ are monic for $n > 0$, of degree $n$ and odd or even for $n$ odd or even respectively. The coefficient of $x^{n-2}$ in $T_n(x)$ is $-n$ for $n \geq 2$.

Corollary 2. $x^n+x^{-n} = T_n(x+x^{-1})$.

Corollary 3. $T_n \circ T_m = T_{nm} = T_m \circ T_m$.

Theorem 7. (Ritt 1922 for $k=\mathbb{C}$). *Let* $m > n > 1$, $(m,n)=1$, *char* $k=0$ *or char* $k > m$. *Then* $G(A) = H(B)$, $G,H,A,B \in k[x]$, $m = |G| = |B|, |H| = |A| = n$, *if and only if there exist linear functions* $L_1, L_2 \in k[x]$ *such that either*

$$(i)\quad \langle L_1 \circ G, A \circ L_2\rangle \sim \langle x^r P(x)^n, x^n\rangle,\ \langle L_1 \circ H, B \circ L_2\rangle \sim \langle x^n, x^r P(x^n)\rangle$$

*for a* $P \in k[x]$ *satisfying* $r+n|P| = m$,

*or*

$$(ii)\quad \langle L_1 \circ G, A \circ L_2\rangle \sim \langle \lambda^{-mn} T_m(\lambda^n x),\ \lambda^{-n} T_n(\lambda x)\rangle,$$

$$\langle L_1 \circ H, B \circ L_2\rangle \sim \langle \lambda^{-mn} T_n(\lambda^m x),\ \lambda^{-m} T_m(\lambda x)\rangle$$

*for a* $\lambda$ *satisfying* $\lambda^2 \in k^*$.

H.Levi 1942 and Dorey and Whaples 1974 proved this theorem for k of characteristic 0, algebraically closed, under the assumption that $G,H,A,B$ are indecomposable. The proof we give below while elementary is long and involved. The reader if he wishes may pass over the proof since the results are not used in the sequel.

Lemma 1. The conditions given in Theorem 7 are sufficient.

Proof.

(i) $L_1 \circ G \circ A \circ L_2 = x^{rn}P(x^n)^n = L_1 \circ H \circ B \circ L_2$, $G \circ A = H \circ B$

(ii) $L_1 \circ G \circ A \circ L_2 = \lambda^{-mn}\, T_m \circ T_n(\lambda x) = \lambda^{-mn}\, T_n \circ T_m(\lambda x) =$

$= L_1 \circ H \circ B \circ L_2$, $G \circ A = H \circ B$.

If $\lambda^2 \in k^*, \lambda^{-n}\, T_n(\lambda x) \in k[x]$ by Corollary 1, hence polynomials $G,H,A,B$ have coefficients in $k$.

Lemma 2. If the condition is necessary for the algebraic closure $\hat{k}$ of $k$ then it is necessary for $k$.

Proof. Case (i). Let

$$G,H,A,B \in k[x],\ \hat{L}_1,\hat{L}_2,\hat{L}_3,\hat{L}_4,\ \hat{P} \in \hat{k}[x];$$

$$\hat{L}_1 \circ G = x^r\hat{P}(x)^n \circ \hat{L}_3, \qquad A \circ \hat{L}_2^{-1} = \hat{L}_3^{-1} \circ x^n$$

$$\hat{L}_1 \circ H = x^n \circ \hat{L}_4, \qquad B \circ \hat{L}_2^{-1} = \hat{L}_4^{-1} \circ x^r\, \hat{P}(x^n);$$

$\hat{L}_i = \lambda_i(x+\mu_i)$, $i \le 4$. First we shall show $\mu_i \in k$. We have

$$\lambda_1(H + \mu_1) = \hat{L}_1 \circ H = \hat{L}_4^n = \lambda_4^n(x + \mu_4)^n.$$

Thus $\lambda_1^{-1}\lambda_4^n$, $n\mu_4$, $\lambda_1^{-1}\lambda_4^n\mu_4^n - \mu_1 \in k$, whence $\mu_4, \mu_1 \in k$. Similarly, from

$$\lambda_3(A + \mu_3) = \hat{L}_3 \circ A = \hat{L}_2^n = \lambda_2^n(x + \mu_2)^n$$

we infer that $\mu_2, \mu_3 \in k$.

Define

$$L_1 = \lambda_1\lambda_4^{-n}(x+\mu_1),\ L_2 = x + \mu_2,\ L_3 = \lambda_3\lambda_2^{-n}(x + \mu_3),$$
$$L_4 = x + \mu_4,\quad P(x) = \lambda_4^{-1}\lambda_2^{r}\ \hat{P}(\lambda_2^{n}x).$$

We get

$$L_4 \circ B \circ L_2^{-1} = \lambda_4^{-1}\ \hat{L}_4 \circ B \circ \hat{L}_2^{-1}(\lambda_2 x) = \lambda_4^{-1}(\lambda_2 x)^{r}\ \hat{P}(\lambda_2^{n}\ x^{n}) =$$
$$x^{r}\ P(x^{n})$$

since $L_2^{-1} = \hat{L}_2^{-1}(\lambda_2 x)$. Hence $P \in k[x]$. Moreover

$$B \circ L_2^{-1} = L_4^{-1} \circ x^{r}P(x^{n}).$$

We check

$$L_1 \circ H = \lambda_4^{-n}\ \hat{L}_1 \circ H = \lambda_4^{-n}\ \hat{L}_4^{n} = (x + \mu_4)^{n} = x^{n} \circ L_4,$$
$$A \circ L_2^{-1} = A \circ \hat{L}_2^{-1}(\lambda_2 x) = \hat{L}_3^{-1} \circ x^{n} \circ (\lambda_2 x)$$
$$= \hat{L}_3^{-1} \circ (\lambda_2 x)^{n} = \lambda_3^{-1}\lambda_2^{n}\ x^{n} - \mu_3 = L_3^{-1} \circ x^{n},$$
$$L_1 \circ G \circ L_3^{-1} \circ x^{n} = L_1 \circ G \circ A \circ L_2^{-1} = L_1 \circ H \circ B \circ L_2^{-1} = x^{rn}\ P(x^{n})^{n},$$
$$L_1 \circ G \circ L_3^{-1} = x^{r}\ P(x)^{n},$$
$$L_1 \circ G = x^{r}\ P(x^{n}) \circ L_3.$$

Hence

$$\langle L_1 \circ G, A \circ L_2^{-1}\rangle \sim \langle x^{r}\ P(x)^{n}, x^{n}\rangle, \langle L_1 \circ H, B \circ L_2^{-1}\rangle \sim \langle x^{n}, x^{r}\ P(x^{n})\rangle.$$

Case (ii). We are given:

$$\hat{L}_1 \circ G = T_m \circ \hat{L}_3, \qquad A \circ \hat{L}_2^{-1} = \hat{L}_3^{-1} \circ T_n,$$
$$\hat{L}_1 \circ H = T_n \circ \hat{L}_4, \qquad B \circ \hat{L}_2^{-1} = \hat{L}_4^{-1} \circ T_m.$$

Let $L_i = \lambda_i(x + \mu_i)$. In the first equation the quotient of the first 2 coefficients on the left is in k: on the right we have $T_m(\lambda_3(x + \mu_3))$ so we get $m\mu_3 \in k$, whence $\mu_3 \in k$. It follows similarly that all $\mu_i \in k$. Let $g_o$ be the leading coefficient of G. From $\hat{L}_1 \circ G = T_n \circ \hat{L}_3$ we get $\lambda_1 g_o = \lambda_3^m$, $\lambda_1\lambda_3^{-m} \in k$. Similarly we have $\lambda_3\lambda_2^{-n} \in k$.
In the identity

$\lambda_1(G+\mu_1) = T_m(\lambda_3(x+\mu_3))$ substitute $x - \mu_3$ for x. We get

$G(x-\mu_3)+\mu_1 = \lambda_1^{-1} T_m(\lambda_3 x)$.

The 3-rd coefficient on the right (see Cor.1) is $-m\lambda_3^{m-2}\lambda_1^{-1} \in k$ thus $\lambda_3^2 \in k$. Similarly $\lambda_2^2 \in k$. We also get

$$\lambda_1^{-1}\lambda_3^m,\ \lambda_3^{-1}\lambda_2^n,\ \lambda_1^{-1}\lambda_2^{mn} \in k.$$

Define $\lambda = \lambda_2$, $L_1 = \lambda_1\lambda_2^{-mn}(x+\mu_1)$, $L_2 = x+\mu_2$,

$$L_3 = \lambda_3\lambda_2^{-n}(x+\mu_3), \quad L_4 = x+\mu_4.$$

We have $\lambda^2 \in k^*$, $L_i \in k[x]$. Equations (ii) hold with $L_2, L_4$ replaced by $L_2^{-1}$, $L_4^{-1}$ respectively.

In the sequel assume k algebraically closed, $c \in k$.

<u>Convention 1</u>. If $D \in k(C)$ let $N_{C/D}$ denote the norm for $k(C)$ to $k(D)$.

<u>Lemma 3</u>. $a_o N_{x/A}(x-c) = (-1)^{|A|-1}(A(x)-A(c))$, where $a_o$ is the leading coefficient of A.

<u>Proof.</u> The conjugates of x are $\xi_i$ satisfying $A(\xi_i)=A(x)$.

Thus

$$a_o N_{x/A}(x-c) = a_o \prod_{i=1}^{|A|} (\xi_i - c) = (-1)^{|A|} a_o \prod_{i=1}^{|A|} (c-\xi_i)$$

$$= (-1)^{|A|}(A(c)-A(x)).$$

Lemma 4. Under the assumption of the theorem,

$$h_o N_{x/A}(B(x)-c) = h_o N_{B/F}(B(x)-c) = (-1)^{|H|-1}(F(x)-H(c)),$$

where $F=G(A) = H(B)$ and $h_o$ is the leading coefficient of $H$.

Proof. It follows from Theorem 5, that

$$[k(A,B):k(B)] = \frac{|B|}{(|A|,|B|)} = |B| = [k(x):k(B)]$$

thus $k(A,B) = k(x)$.

Moreover

$$N_{k(A,B)/k(A)}(B(x)-c) \mid N_{k(B)/k(A)\cap k(B)}(B(x)-c).$$

Since the polynomials are of the same degree on both sides they must coincide and by Lemma 3

$$h_o N_{k(A,B)/k(A)}(B(x)-c) = h_o N_{k(B)/k(F)}(B(x)-c)$$

$$= (-1)^{|H|-1}(F(x)-H(c)).$$

Lemma 5. If $H$ takes only one value $h$ for all zeros of $H'$

then the case (i) of the theorem holds. The same holds for G.

Proof. If $H' = c\Pi(x-\beta_i)^{\mu_i}$, then

$$H = h + \Pi(x - \beta_i)^{\rho_i} U(x), \quad (x - \beta_i, U) = 1 \text{ for all } i.$$

But then

$$H' = \Pi(x - \beta_i)^{\rho_i - 1} V(x), \quad (x - \beta_i, V) = 1 \text{ for all } i,$$

whence

$$H' = c(x - \beta)^{\mu}, \quad H = h + \frac{c}{\mu+1}(x - \beta)^{\mu+1}.$$

Therefore $H(x) = L_1^{-1} \circ x^n \circ L_4^{-1}$. It follows that $k(x)$ is cyclic over $k(H)$, whence $k(B)$ is cyclic over $k(H(B)) = k(G(A))$ and $k(A,B)$ is cyclic over $k(A,G(A)) = k(A)$. But $k(A,B) = k(x)$, hence the equation $A(z)-A = 0$ for $z$ is cyclic with respect to $k(A)$. Since all of its zeros lie in $k(x)$ they are rational functions of $x$. If $g(x)$ is a zero then $A(g(x)) = A$, whence $g(x)$ is a polynomial and must be linear. Thus all roots of $A(z)-A = 0$ are of form $dx+e$. However, as already noted $k(x)_{/k(A)}$ is cyclic. Hence one substitution $S$ generates the Galois group.

If $S(x) = dx+e$ we have $S^i(x) = d^i x + e\frac{d^i-1}{d-1}$ $(d \neq 1)$. Since $|A| = n$ it follows that $d^n = 1$. Furthermore

$$S(x - \frac{e}{1-d}) = dx+e - \frac{e}{1-d} = dx - \frac{ed}{1-d} = d(x - \frac{e}{1-d}),$$

so that

$$S((x - \tfrac{e}{1-d})^n) = d^n(x - \tfrac{e}{1-d})^n = (x - \tfrac{e}{1-d})^n,$$

whence $C = (x - \frac{e}{1-d})^n \in k(A)$.

Since $C$ and $A$ have the same degree,

$A = L_3 \circ (x - \frac{e}{1-d})^n = L_3 \circ x^n \circ L_2^{-1}$ for appropriate $L_2$ and $L_3$.

Then

$$G(A) = H(B) \quad \text{implies}$$
$$G \circ L_3 \circ x^n = L_1^{-1} \circ x^n \circ L_4^{-1} \circ B \circ L_2,$$

and

$$L_1 \circ G \circ L_3 \circ x^n = (L_4^{-1} \circ B \circ L_2)^n.$$

Let $L_4^{-1} \circ B \circ L_2 = x^r I(x)$, where $I(0) \neq 0$. Then $L_1 \circ G \circ L_3 \circ x^n = x^{rn}\ I(x)^n$. This last equation shows the substitution $x \longrightarrow \zeta_n x$ does not change $I(x)^n$. But $I(\zeta_n x) \neq \zeta_n^r\ I(x)$ for all $r \not\equiv 0 \mod n$ since $I(0) \neq 0$. Hence $I(\zeta_n x) = I(x)$ and

$$I(x) = P(x^n).$$

Thus $L_1 \circ G \circ L_3 \circ x^n = (x^n)^r\ P(x^n)^n$ and we have

$$L_1 \circ G \circ L_3 = x^r\ P(x)^n.$$

Thus we have established case (i) to hold under the hypothesis of the lemma about H.

If $G$ takes only one value for all the zeros of $G'$ then by the same argument we obtain

$$G(x) = \tilde{L}_1^{-1} \circ x^m \circ \tilde{L}_4^{-1}, \quad B(x) = \tilde{L}_3 \circ x^m \circ \tilde{L}_2^{-1},$$
$$\tilde{L}_4^{-1} \circ A \circ \tilde{L}_2 = x^{\tilde{r}} \tilde{P}(x^m), \quad \tilde{L}_1 \circ H \circ \tilde{L}_3 = x^{\tilde{r}} \tilde{P}(x)^n.$$

Since however $|A| = |H| = n < m$ we must have $|\tilde{P}| = 0$, $\tilde{P} = c \in k$, $\tilde{r} = n$. Taking $P = 1$, $r = m$, $L_1 = c^{-m} \tilde{L}_1$, $L_2 = \tilde{L}_2$ we get again (i).

Now we need no longer assume that $m > n$, but only that char $k > \max\{|m|, |n|\}$ or char $k = 0$.

Lemma 6. If char $k \neq 2$ the equation

$$(Q(x)-q_1)(Q(x)-q_2) = (x-\xi_1)(x-\xi_2)R^2(x), \quad Q, R \in k[x],$$
$$q_1, q_2, \xi_1, \xi_2 \in k, \quad q_1 \neq q_2, \quad \xi_1 \neq \xi_2$$

implies

$$Q(x) = L \circ T_{|Q|} \circ M^{-1},$$

where $L, M$ are linear functions with coefficients in $k$.

Proof. (following Dorey and Whaples 1974). Without loss of generality we can assume that either

$$(1) \qquad Q(x) - q_i = (x - \xi_i) R_i^2(x) \qquad (i = 1,2)$$

or

$$(2) \qquad Q(x) - q_1 = (x - \xi_1)(x - \xi_2) R_3^2(x), \quad Q(x) - q_2 = R_4(x)^2,$$

where $R_i(x) \in \hat{k}[x]$. Put

$$P(x) = L^{-1} \circ Q \circ M,$$

where

$$L(x) = \frac{q_1-q_2}{4} x + \frac{q_1+q_2}{2}, \quad M(x) = \frac{\xi_1-\xi_2}{4} x + \frac{\xi_1+\xi_2}{2}.$$

In case (1) we get

$$(3) \qquad \frac{q_1-q_2}{4} (P(x) \pm 2) = \frac{\xi_1-\xi_2}{4} (x \pm 2) \; R^2_{\frac{3\pm 1}{2}} (M(x))$$

and in case (2)

$$(4) \qquad \frac{q_1-q_2}{4} (P(x)-2) = \left(\frac{\xi_1-\xi_2}{4}\right)^2 (x^2-4) \; R_3^2(M(x)),$$

$$\frac{q_1-q_2}{4} (P(x))+2) = R_4(M(x))^2.$$

Substitute now $x = z+z^{-1}$ in (3) and (4). In each instance we obtain

$$P(z+z^{-1})-2 = z^{-|P|} S_1(z)^2, \; P(z+z^{-1})+2 = z^{-|P|} S_2(z)^2$$

where

$$S_1, S_2 \in \hat{k}[z], \; |S_1| = |S_2| = |P| \text{ and } S_1(1) = 0. \text{ Thus}$$

$$4z^{|P|} = S_2^2 - S_1^2 = (S_2-S_1)(S_2+S_1).$$

Since char $k \neq 2$, $\max(|S_2-S_1|, |S_2+S_1|) = |P|$, hence $\min(|S_2-S_1|, |S_2+S_1|) = 0$ and we can assume that $S_2-S_1 = s \in \hat{k}$.

Then $$s(2S_1 + s) = 4z^{|P|}$$

and on substituting $z=1$ we get $s^2=4$. Now

$$S_1 = \frac{4z^{|P|}-s^2}{2s} = \frac{2}{s}(z^{|P|}-1)$$

and

$$P(z+z^{-1}) = 2+z^{-|P|}S_1(z)^2 = 2+\frac{4}{s^2}z^{-|P|}(z^{|P|}-1)^2 = z^{|P|}+z^{-|P|},$$

Hence by Cor.2 to Definition 6 $P(x) = T_{|P|}(x)$, which proves the lemma since $|P|=|Q|$.

Lemma 7. Let $Q(x) \in k[x]$, $q,q' \in k$, $q \neq q'$, $|Q| \not\equiv 0 \bmod \operatorname{char} k$. If the polynomial $(Q-q)(Q-q')$ has at most two simple zeros then it has exactly two, say $b,b'$, all the remaining zeros are double. Furthermore, we then have

$$(5) \qquad |Q|^2(Q-q)(Q-q') = (x-b)(x-b')Q'^2$$

and

$$(6) \qquad N_{x/Q}\, Q'(x) = c(Q-q)^{\frac{|Q|-1-\varepsilon}{2}}(Q-q')^{\frac{|Q|-1+\varepsilon}{2}}$$

where $1+\varepsilon$ is the number of simple zeros of $Q-q$ and $c \in k$.

Proof. Let

$$Q(x)-q \overset{\text{can}}{\underset{\bar{k}}{=}} a_o \prod_{i=1}^{r}(x-b_i)^{\alpha_i},$$

$$Q(x)-q' \overset{\text{can}}{\underset{\bar{k}}{=}} a_o \prod_{j=1}^{s}(x-b'_j)^{\alpha'_j}.$$

We note that since $q \neq q'$ no $b_i$ is equal to a $b'_j$.

We have

$$\prod_{i=1}^{r} (x-b_i)^{\alpha_i - 1} \prod_{j=1}^{s} (x-b'_j)^{\alpha'_j - 1} \,|\, Q'(x),$$

whence

$$|Q|-1 = |Q'| \geq \sum_{i=1}^{r} (\alpha_i - 1) + \sum_{j=1}^{s} (\alpha'_j - 1) = 2|Q|-r-s,$$

$$r+s-1 \geq |Q| = \sum_{i=1}^{r} \alpha_i = \sum_{j=1}^{s} \alpha'_j,$$

and $$2r+2s-2 \geq \sum_{i=1}^{r} \alpha_i + \sum_{j=1}^{s} \alpha'_j.$$

Since all exponents $\alpha_i, \alpha'_j$ except at most two are at least 2, exactly two are equal to 1 and the others equal 2. Hence up to a permutation either

$$\alpha_1 = \alpha'_1 = 1, \quad \alpha_i = \alpha'_j = 2 \quad \text{for } i \geq 2,\ j \geq 2,$$

or

$$\alpha_1 = \alpha_2 = 1, \quad \alpha_i = \alpha'_j = 2 \quad \text{for } i > 2,\ j > 0,$$

or

$$\alpha'_1 = \alpha'_2 = 1, \quad \alpha_i = \alpha'_j = 2 \quad \text{for } i > 0,\ j > 2.$$

In any case

$$Q' = a_0 |Q| \prod_{i=2+\varepsilon}^{r} (x-b_i) \prod_{j=2-\varepsilon}^{s} (x-b'_j), \quad r = \frac{|Q|-1+\varepsilon}{2}, s = \frac{|Q|-1-\varepsilon}{2}.$$

The formula (5) follows at once. To prove (6) we use Lemma 3. We get

$$N_{x/Q}(Q') = \prod_{i=2+\varepsilon}^{r} (Q(x) - Q(b_j)) \prod_{j=2-\varepsilon}^{s} (Q(x)-Q(b_j'))$$

$$= c(Q-q)^{\frac{|Q|-1-\varepsilon}{2}} (Q-q')^{\frac{|Q|-1+\varepsilon}{2}} .$$

_Lemma 8_. If $H-h$, $H-h'$ $(h \neq h')$ each have at most one simple zero then the case (ii) of Theorem 7 holds.

_Proof_. By Lemma 7 $H-h$, $H-h'$ each have exactly one simple zero and the remaining zeros double. Moreover, since $|H| = n$

$$(7) \qquad n^2(H-h)(H-h') = (x-b)(x-b')H'^2$$

and

$$N_{x/H}(H') = c_0(H-h)^{\frac{n-1}{2}}(H-h')^{\frac{n-1}{2}} .$$

On the other hand, the equation $G(A) = H(B)$ implies

$$(8) \qquad G'(A)A' = H'(B)B'.$$

Following Levi we take norms from $k(x)$ to $k(A(x))$ and get

$$N_{x/A}G'(A)N_{x/A}A' = N_{x/A}H'(B)N_{x/A}B',$$

whence in virtue of Lemma 4

(9) $$G'(A)^n N_{x/A}A' = N_{B/F}H'(B)N_{x/A}B' = c_1(H(B)-h)^{\frac{n-1}{2}}(H(B)-h')^{\frac{n-1}{2}}N_{x/A}B' =$$

$$c_1(G(A)-h)^{\frac{n-1}{2}}(G(A)-h')^{\frac{n-1}{2}}N_{x/A}B'.$$

By Lemma 7 $(G-h)(G-h')$ has at least two simple zeros, hence $(G(A)-h)(G(A)-h')$ has two factors $A-a, A-a'$ prime to $G'(A)$. Suppose there are three such factors. On the right hand side of (9) we get $\frac{3}{2}(n-1)$ such factors each of degree $n$ all dividing $N_{x/A}A'$, which has degree $n(n-1)$ by Lemma 3, a contradiction.

We must have precisely two factors in question, therefore, $(G-h)(G-h')$ has precisely two simple zeros. Furthermore by Lemma 7 all zeros of $(G-h)(G-h')$ except $a, a'$ are double

(10) $$N_{x/G}G'(x) = c_2(G-h)^{\frac{m-1-\varepsilon}{2}}(G-h')^{\frac{m-1+\varepsilon}{2}} \qquad (\varepsilon=0,\pm 1)$$

and

(11) $$m^2(G-h)(G-h') = (x-a)(x-a')G'^2.$$

Hence, $m^2(G(A)-h)(G(A)-h') = (A-a)(A-a')G'(A)^2$ and

$$G'(A)N_{x/A}A' = c_3(A-a)^{\frac{n-1}{2}}(A-a')^{\frac{n-1}{2}}N_{x/A}B'.$$

On comparing the degrees we get

$$G'(A) = c_4 N_{x/A}B',$$

$$N_{x/A}A' = c_5(A-a)^{\frac{n-1}{2}}(A-a')^{\frac{n-1}{2}}.$$

It follows that B' has only simple zeros. Indeed, by Lemma 3

$$B' = c_6 \Pi (x - \xi_i)^{\alpha_i} \quad \text{implies} \quad N_{x/A} B' = c_7 \Pi (A(x) - A(\xi_i))^{\alpha_i} = G'(A)$$

thus

$$c_7 \Pi (x - A(\xi_i))^{\alpha_i} = G'(x)$$

and since by (11) $G'$ has only simple zeros all $\alpha_i$ are 1.

On the other hand taking norms of both sides of (8) from $k(x)$ to $k(B(x))$ we get

$$N_{x/B} G'(A) N_{x/B} A' = N_{x/B} H'(B) \; N_{x/B} B'$$

and by virtue of Lemma 4

$$N_{A/F} G'(A) N_{x/B} A' = H'(B)^m N_{x/B} B'.$$

Hence by (10) we have

$$c_2 (F - h)^{\frac{m-1-\varepsilon}{2}} (F - h')^{\frac{m-1+\varepsilon}{2}} N_{x/B} A' = H'(B)^m N_{x/B} B'.$$

However F-h has a factor B-b prime to $H'(B)$ and similarly $F - h'$. Hence

$$(B - b)^{\frac{m-1-\varepsilon}{2}} (B - b')^{\frac{m-1+\varepsilon}{2}} = c_8 N_{x/B} B'.$$

It follows that $B - b$ has $\frac{m-1-\varepsilon}{2}$ double zeros, $B - b'$ has $\frac{m-1+\varepsilon}{2}$ double zeros and $(B - b)(B - b')$ has exactly two simple zeros $\xi, \xi'$.

By Lemma 7

$$m^2(B-b)(B-b') = (x-\xi)(x-\xi')B'^2$$

Hence by (7)

$$m^2n^2(F-h)(F-h') = m^2(B-b)(B-b')H'(B)^2 =$$

$$(x-\xi)(x-\xi')(H'(B)B')^2.$$

It follows by Lemma 6 that

$$F = L \circ T_{mn} \circ M^{-1} = L \circ T_m \circ T_n \circ M^{-1} = L \circ T_n \circ T_m \circ M^{-1}.$$

On the other hand

$$F = G \circ A = H \circ B$$

and by Theorem 5 $k(A) = k(T_n \circ M^{-1})$, $k(B) = k(T_n \circ M^{-1})$.

Hence we get $A = L_3 \circ T_n \circ M^{-1}$, $B = L_4 \circ T_m \circ M^{-1}$

$$G = L \circ T_m \circ L_3^{-1}, \quad H = L \circ T_n \circ L_4^{-1}$$

i.e. $\langle L^{-1} \circ G, A \circ M\rangle \sim \langle T_m, T_n\rangle$,

$$\langle L^{-1} \circ H, B \circ M\rangle \sim \langle T_n, T_m\rangle$$

and the case (ii) holds with $L_1 = L^{-1}$, $L_2 = M$.

Convention 2 If $a \in k$ and $P \in k[x]$, define

$$\operatorname{ind}_a P = |(P', P-a)|.$$

Lemma 9. If $\deg P < \operatorname{char} k$ or $\operatorname{char} k = 0$, then

$$\sum_{a \in k} \operatorname{ind}_a P = |P| - 1.$$

Proof. Observe that $\operatorname{ind}_a P \neq 0$ only if $a = P(\xi)$, for some zero $\xi$ of $P'$. Hence if $P' \underset{k}{\overset{can}{=}} \text{const} \prod_{i=1}^{j} (x - \xi_i)^{\alpha_i}$

then $\sum_a \operatorname{ind}_a P = \sum_{i=1}^{j} \operatorname{ind}_{P(\xi_i)} P$. But

$(x - \xi_i)^{\alpha_i + 1} \| P(x) - P(\xi_i)$ thus $\sum_a \operatorname{ind}_a P = \sum_{i=1}^{j} \alpha_i = |P| - 1.$

Lemma 10. Let A,B,G,H as in the Theorem. If $H(y) - G(a) \underset{k}{\overset{can}{=}}$

$\text{const} \prod_{q=1}^{s} (y - \eta_q)^{j_q}$, and $(x-a)^{\mu} \| G(x) - G(a)$,

then $\operatorname{ind}_{G(a)} H \geq \operatorname{ind}_a A \geq \sum_{q=1}^{s} (j_q - (j_q, \mu))$, with equality holding if $\mu = 1$.

Proof. Let $A(x) - a \underset{k}{\overset{can}{=}} \text{const} \prod_{p=1}^{r} (x - \xi_p)^{i_p}$. Then, by

Lemma 3, $N_{x/B}(A(x)-a) = \text{const} \prod_{p=1}^{r} N_{x/B}(x - \xi_p)^{i_p} =$

$\text{const} \prod_{p=1}^{r} (B(x) - B(\xi_p))^{i_p}$ and by Lemma 4, $N_{x/B}(A(x)-a) =$

$N_{A/F}(A(x)-a) = \text{const}\ (F - G(a)) = \text{const}\ (H(B) - G(a))$.

Hence $H(y) - G(a) = c \prod_{p=1}^{r} (y - B(\xi_p))^{i_p}$ and $j_q = \sum_{B(\xi_p) = \eta_q} i_p$

for $q \leq s$.

Let $\mu_p$ be the multiplicity of $\xi_p$ as zero of $B(x)-B(\xi_p)$. It follows from $(x-\xi_p)^{i_p} \| A(x)-a$, $(A(x)-a)^{\mu} \| F-G(a)$, that

$$\mu i_p = \mu_p j_q, \quad \text{whenever} \quad \eta_q = B(\xi_p). \quad \text{Thus}$$

$$i_p \equiv 0 \left(\bmod \frac{j_q}{(j_q,\mu)}\right) \quad \text{and} \quad \sum_{B(\xi_p)=\eta_q} 1 \leq (j_q,\mu).$$

Hence

$$\operatorname{ind}_{G(a)} H = \sum_{q=1}^{s} (j_q - 1) = \sum_{q=1}^{s} j_q - s = \sum_{p=1}^{r} i_p - s = \sum_{p=1}^{r} (i_p - 1) + r - s =$$

$$= \operatorname{ind}_a A + r - s \geq \operatorname{ind}_a A = \sum_{q=1}^{s} \sum_{\substack{p=1 \\ B(\xi_p)=\eta_q}}^{r} (i_p - 1) = \sum_q \sum_{B(\xi_p)=\eta_q} i_p - \sum_{B(\xi_p)=\eta_q} 1$$

$$\geq \sum_{q=1}^{s} (j_q - (j_q,\mu)).$$

If $\mu = 1$ we have equality everywhere.

Convention 3. $h$ is an extra point of $H$ if and only if

$$\operatorname{ind}_h H > \sum_{G(a)=h} \operatorname{ind}_a A.$$

Extra points of $G$ are defined similarly.

Lemma 11. If $h$ is an extra point of $H$, then $G-h$ has no simple zero.

Proof. If $G-h$ has a simple zero $a$ then we apply Lemma 10 with $\mu = 1$ and get $\operatorname{ind}_h H = \operatorname{ind}_a A$, contrary to the definition of an extra point.

Lemma 12. $G$ and $H$ cannot each have an extra point.

Proof. Suppose that $g$ is an extra point of $G$ and $h$ is an extra point of $H$.
If $g \neq h$ then by Lemma 11 $G-h$, $H-g$ have no simple zeros, hence $F-h$, $F-g$ have no simple zero. Consequently $F'$ would have $\frac{|F|}{2}$ zeros coming from $F-h$ and $\frac{|F|}{2}$ zeros coming from $F-g$, together $|F|$ zeros, a contradiction.
There remains the possibility of a common extra point $e$. Let

$$G(x) - e \overset{\text{can}}{\underset{k}{=}} \text{const} \prod_{p=1}^{r} (x-a_p)^{i_p},$$

$$H(x) - e \overset{\text{can}}{\underset{k}{=}} \text{const} \prod_{q=1}^{s} (x-b_q)^{j_q}.$$

By Lemma 10

$$\sum_{p=1}^{r} \text{ind}_{a_p} A \geq \sum_{p=1}^{r} \left( \sum_{q=1}^{s} (j_q - (j_q, i_p)\right)$$

$$= \sum_{q=1}^{s} \sum_{p=1}^{r} (j_q - (j_q, i_p)).$$

Suppose for every $q$, that either there are two $i_p$ not divisible by $j_q$ or there is an $i_p$ prime to $j_q$. We get
$\sum_{p=1}^{r} (j_q - (j_q, i_p)) \geq \frac{j_q}{2} + \frac{j_q}{2} = j_q$ or $j_q - 1$, so
$\sum_{q=1}^{s} \sum_{p=1}^{r} (j_q - (j_q, i_p)) \geq \sum_{j=1}^{s} (j_q - 1) = \text{ind}_e H$, contrary to the assumption that $e$ is a common extra point.
We may suppose that there is at most one $i_p$ not divisible by $j_1$, and this $i_p$ satisfies $(i_p, j_1) = d > 1$. Hence $d | i_p$ for all $p$, $|G| \equiv 0 \pmod d$. But $(|G|, |H|) = 1$, thus $d \nmid |H|$, so $d$ does not divide all $j_q$ and so for some $q$ we have

(12) $\quad j_q \not\equiv 0 \pmod{i_p}$ for $1 \leq p \leq r$.

Applying the same inequality with $B$ instead of $A$ we obtain

$$S = \sum_{q=1}^{s} \operatorname{ind}_{b_q} B \geq \sum_{q=1}^{s} \sum_{p=1}^{r} (i_p - (i_p, j_q)).$$

By (12) $S \geq \sum_{p=1}^{r} \frac{i_p}{2} = \frac{m}{2}$. By symmetry, also $\sum_{p=1}^{r} \operatorname{ind}_{a_p} A \geq \frac{n}{2}$.

But since $e$ is a common extra point of $G$ and $H$ we have

$\operatorname{ind}_e H = \frac{n}{2} + \nu \; (\nu > 0)$, $\operatorname{ind}_e G = \frac{m}{2} + \mu \; (\mu > 0)$. We can assume $\mu \geq \nu$.

Since

$$|H| - 1 = n - 1 \geq \operatorname{ind}_e H > \sum_{G(a)=e} \operatorname{ind}_a A,$$

and by Lemma 9 $\sum_{a \in k} \operatorname{ind}_a A = |A| - 1 = n - 1$ there is a $c$ such that $\operatorname{ind}_c A > 0$ and $G(c) \neq e$.

By the same lemma $\operatorname{ind}_{G(c)} G + \operatorname{ind}_e G \leq m - 1$, whence

$$\operatorname{ind}_{G(c)} G \leq m - 1 - \left(\frac{m}{2} + \mu\right) \leq \frac{m - (2\mu + 2)}{2}$$

and thus $G(x) - G(c)$ has at least $2\mu+2$ simple zeros.

If $a_1', a_2', \ldots, a_{2\mu+2}'$ are the simple zeros in question, then by Lemma 10

$$\operatorname{ind}_{a_\lambda'} A = \operatorname{ind}_{G(c)} H \geq \operatorname{ind}_c A \geq 1.$$

Hence

$$\sum_{\lambda=1}^{2\mu+2} \operatorname{ind}_{a_\lambda'} A = (2\mu+2) \operatorname{ind}_{G(c)} H \geq \operatorname{ind}_{G(c)} H + (2\mu+1)$$

and

$$\sum_{G(a)=G(c)} \operatorname{ind}_a A \geq \operatorname{ind}_{G(c)} H+2\mu+1.$$

We also have

$$\sum_{G(a)=e} \operatorname{ind}_a A \geq \frac{n}{2} = \operatorname{ind}_e H - \nu.$$

For any point $h \neq G(c), e$

$$\sum_{G(a)=h} \operatorname{ind}_a A \geq \operatorname{ind}_h H.$$

Now sum the last 3 inequalities and use Lemma 9:

$$n-1 \geq n-1-\nu+2\mu+1 > n-1, \quad \text{a contradiction.}$$

Proof of Theorem 7. By Lemma 12 $G$ and $H$ cannot both have an extra point. Suppose $G$ has no extra point. If there is at most one $g$ such that $\operatorname{ind}_g G > 0$ Lemma 5 applies and case (i) holds. Suppose there are at least two $g$ such that $\operatorname{ind}_g G > 0$. For each such $g$, $H-g$ must have at most one simple zero. For if $b$ is a simple zero of $H-g$, then by Lemma 10 we have $\operatorname{ind}_b B = \operatorname{ind}_g G$. Hence if $H-g$ has at least 2 simple zeros then

$$\sum_{H(b)=g} \operatorname{ind}_b B \geq 2 \operatorname{ind}_g G > \operatorname{ind}_g G.$$

But then, by Lemma 9 there exists a point $e$ such that

$\sum_{H(b)=e} \operatorname{ind}_b B < \operatorname{ind}_e G$ i.e. $e$ is an extra point of $G$, contrary to our assumption on $G$. Thus there exist $g_1, g_2$

such that $H - g_1$, $H - g_2$ each have at most one simple zero and Lemma 8 applies and case (ii) holds.

This completes the proof of Ritt's second theorem.

Remarks 1. Using the approach of Dorey and Whaples which extends to the case $\operatorname{char} k > \max\{m,n\}$ it is possible to give a more advanced but shorter proof of Lemma 8.

2. It is not known to the writer how essential is the assumption $\operatorname{char} k > \max\{m,n\}$. If $p = \operatorname{char} k$ and $S$ is the Frobenius endomorphism $a \longrightarrow a^p$ of $k$ we have for every $F \in k[x]$, $x^p \circ F = SF \circ x^p$, thus the theorem fails if $mn \equiv 0 \mod \operatorname{char} k$ and $k$ is infinite.

## Section 6. Applications to Diophantine equations

Theorem E. (Siegel 1929) *If an equation* $F(x,y) = 0$ *where* $F \in Q[x,y]$ *has infinitely many rational solutions with bounded denominators, then there exist* $p,q \in Q(t)$ *such that* $F(p(t),q(t)) = 0$, $p,q$ *not both constant.*

Theorem 8. *Let* $H,G \in \mathbb{Q}[x]$, $|H| = n$, $|G| = m$, $(n,m) = 1$, $m > n \geq 1$. *The equation* $G(x) = H(y)$ *has infinitely many rational solutions with bounded denominators if and only if either*

$$\text{(i)} \quad G = L_1 \circ x^r P(x)^n \circ L_2, \quad H = L_1 \circ x^n \circ L_3,$$

*or*

$$\text{(ii)} \quad G = L_1 \circ \lambda^{-mn} T_m(\lambda^n L_2), \quad H = L_1 \circ \lambda^{-mn} T_n(\lambda^m L_3)$$

*for suitable linear functions* $L_1, L_2, L_3$, *a polynomial* $P \in \mathbb{Q}[x]$, *and a* $\lambda$ *such that* $\lambda^2 \in \mathbb{Q}$.

Lemma. If $f(x) = H(b(x)) = G(a(x))$, $a,b \in k(x)$, under the assumption of Theorem 7, then there exist polynomials $A,B \in k[x]$ and $\ell \in k(x)$ such that $a = A(\ell)$, $b = B(\ell)$, $|A| = n$, $|B| = m$.

Proof. $k(a,b) = k(\ell)$ by Lüroth's theorem. Moreover

$$[k(a) : k(f)] = |G| = m,$$

$$[k(b) : k(f)] = |H| = n,$$

$$[k(a,b) : k(f)] = mn,$$

$$[k(\ell) : k(a)] = n,$$

$$[k(\ell) : k(b)] = m.$$

Let $a = \frac{A_1(\ell)}{A_2(\ell)}$, $b = \frac{B_1(\ell)}{B_2(\ell)}$, $(A_1,A_2) = (B_1,B_2) = 1$. Thus

$$\max\{|A_1|,|A_2|\} = n,$$

$$\max\{|B_1|,|B_2|\} = m,$$

$$H\left(\frac{B_1(\ell)}{B_2(\ell)}\right) = G\left(\frac{A_1(\ell)}{A_2(\ell)}\right).$$

Compare denominators on both sides. We get $B_2^n = cA_2^m$, and since $(m,n) = 1$ there exists a polynomial $C$ such that

$$B_2 = dC^m, \quad A_2 = eC^n, \quad c,d,e \in k.$$

Either $C$ is constant and the lemma follows, or $C$ is linear, $C(\ell) = \gamma\ell + \delta$. Substitute $\ell = \frac{t^{-1}-\delta}{\gamma}$. Then

$$a = \frac{A_1\left(\frac{t^{-1}-\delta}{\gamma}\right)}{e(t^{-1})^n} = \frac{t^n}{e} A_1\left(\frac{t^{-1}-\delta}{\gamma}\right), \quad b = \frac{t^m}{d} B_1\left(\frac{t^{-1}-\delta}{\gamma}\right)$$

and clearly $a,b$ are polynomials in $t$. This proves the lemma.

<u>Proof of Theorem 8</u>. <u>Necessity</u>. By Theorem E, $G(a)-H(b)=0$, $a,b \in \mathbb{Q}(x)$ not both constant. By the lemma, we get $G(A) = H(B)$. By Theorem 7 we have either case (i) or case (ii).

<u>Sufficiency</u>. In case (i) or (ii) we have $G(L_2^{-1} \circ t^n) = H(L_3^{-1} \circ t^r P(t^n))$ or

$$G(L_2^{-1} \circ \lambda^{-n} T_n(t\lambda)) = H(L_3^{-1} \circ \lambda^{-m} T_m(\lambda t)), \quad \text{respectively.}$$

When $t$ runs through integers, the arguments of $G$ and $H$ have bounded denominators.

The assumption $(|G|, |H|) = 1$ is essential, as the following example shows $G = x^2$, $H = 2y^2 + 1$. Neither the lemma nor the theorem is true in this case. More general but less precise results can be found in Fried 1974a.

Definition 7. *If* $F \in k[x_1,\ldots,x_n]$, $|F| = \max_{1\le i\le n} |F|_{x_i} < d$ *then* $S_d : F \longrightarrow F(x, x^d, x^{d^2}, \ldots, x^{d^{n-1}})$ *is called Kronecker's substitution.*

Corollary 1. If $F_o \in k[x]$ satisfies $|F_o| \le d^n - 1$, then there exists a unique $F \in k[x_1,\ldots,x_n]$ with $|F| < d$ such that

$$S_d F = F_o.$$

Proof. Every non-negative integer $m$ is uniquely represented as $\sum_{j=1}^{n} c_j d^{j-1}$, where $0 \le c_j = c_j(m) < d$.

If $F_o = \sum_m a_m x^m$, then

$$F = \sum_m a_m x_1^{c_1(m)} \ldots x_n^{c_n(m)}.$$

Corollary 2. (Kronecker 1882). $F(x_1,\ldots,x_n)$ with $|F| < d$ is irreducible over $k$ if and only if for every factorization $S_d F = S_d G \cdot S_d H$ with $G, H \in k[x_1,\ldots,x_n]$, $|G|, |H| < d$, we have $F \neq G \cdot H$.

Proof. The necessity of the condition is obvious and the sufficiency follows from the identity $S_d GH = S_d G \cdot S_d H$.

Definition 8. *An integral polynomial is a polynomial with coefficients in* $Z$.

Remark. Integral polynomials are defined also over fields of positive characteristic, since $n = 1 + 1 + \ldots + 1$ (n times).

Convention. A complete polynomial $A(x_1,\ldots,x_n)$ with $|A| = \alpha$ and indeterminate coefficients is a polynomial

$$\sum_{i_1=0}^{\alpha} \sum_{i_2=0}^{\alpha} \cdots \sum_{i_n=0}^{\alpha} a_{i_1 i_2 \ldots i_n} x_1^{i_1} x_2^{i_2} \ldots x_n^{i_n},$$

where $a_{i_1 i_2 \ldots i_n}$ are indeterminates.

Theorem 9. *Let* $B(\mathbf{x}) = A_1(\mathbf{x})\ldots A_l(\mathbf{x})$ *where each* $A_i(\mathbf{x})$ *is a complete polynomial in* $\mathbf{x}$ *with given* $|A_i|$ *and indeterminate coefficients.*

*There exists a non-empty set of non-zero integral forms* $F_j$ $(1 \le j \le \kappa)$ *in the coefficients of* $A_1,\ldots,A_{l-1}$ *such that every integral multilinear form* $M$ *in the coefficients of* $A_1,\ldots,A_l$ *satisfies identities:*

$$(1) \qquad MF_i = \sum_{j=1}^{k} b_{ijM} F_j \qquad (1 \le i \le \kappa),$$

*where* $b_{ijM}$ *are integral linear forms in the coefficients of* $B$*. When the coefficients of* $A_i$ *are specialized at least one* $F_i$ *is non-zero, unless* $A_1 A_2 \cdot \ldots \cdot A_{l-1} = 0$ *for that specialization.*

Proof. (following Prüfer 1932). 1 st step $\mathbf{x} = \langle x \rangle$, $\ell = 2$

$$B(x) = A_1(x)A_2(x), \quad \text{where} \quad A_\nu(x) = \sum_j a_j^{(\nu)} x^j \text{ for } \nu=1,2.$$

Let $n = |A_2|$. For the $F_i$ we take all monic monomials of degree $n$ in the coefficients of $A_1$. In order to prove (1) is sufficient to consider $M = a_i^{(1)} a_j^{(2)}$. Write any product of $a_j^{(2)}$ by a monomial of degree $n+1$ in the $a_j^{(1)}$s as

$$(2) \qquad \underbrace{a_r^{(1)} a_s^{(1)} \ldots a_j}_{j+1 \text{ terms}}{}^{(2)} a_t^{(1)} \ldots,$$

where $r \leq s \leq \ldots \leq t \leq \ldots$ Now we order all products of the form (2) lexicographically, the order of letters being:

$$a_o^{(2)}, a_1^{(2)}, a_2^{(2)}, \ldots, a_n^{(2)}, a_o^{(1)}, a_1^{(1)}, \ldots, a_n^{(1)}.$$

The first product of type (2) in the lexicographic order is

$$a_o^{(1)} a_o^{(2)} a_o^{(1)^n}$$

and it is expressible as $b_o F_1$, where $b_o$ is the leading coefficient of $B$, $F_1 = a_o^{(1)^n}$. We now proceed by induction with respect to the place occupied by the product (2) in the sequence of all products ordered lexicographically. Suppose the assertion is true for all products coming before

$$P = \underbrace{\ldots a_r^{(1)} a_s^{(1)}}_{j+1} a_j^{(2)} a_t^{(1)} \ldots, \text{ where } j > 0, \quad r \leq s \leq t.$$

We have

$$b_{s+j} = \ldots + a_{s-1}^{(1)} a_{j+1}^{(2)} + a_s^{(1)} a_j^{(2)} + a_{s+1}^{(1)} a_{j-1}^{(2)} + \ldots$$

Multiply both sides by the product $F_\mu$ of all factors of $P$ except $a_s^{(1)} a_j^{(2)}$. We get

$$b_{s+j} F_\mu = \ldots a_r^{(1)} a_{s-1}^{(1)} a_{j+1}^{(2)} a_t^{(1)} \ldots + P + \ldots a_r^{(1)} a_{j-1}^{(2)} a_{s+1}^{(1)} a_t^{(1)} \ldots + \ldots$$

and now change the places of factors to keep the products written as in (2). All other terms on the right hand side precede $P$ in the lexicographic order and the inductive assumption applies to them. Formula (1) follows by induction. The statement about specialization is clear from the construction of $F_i$.

<u>2 nd step</u> $\mathbf{x} = \langle x \rangle$, $\ell > 2$. Suppose the theorem true for $\ell-1$ polynomials and let

$$B(x) = A_1(x)\cdot\ldots\cdot A_\ell(x),$$

where $A_i$ are complete polynomials with given $|A_i|$ and indeterminate coefficients.

By the inductive assumption, there is a non-empty finite set of non-zero integral forms $F_j^{(\ell-1)}$ $(1 \le j \le \kappa)$ in the coefficients of $A_1, A_2, \ldots, A_{\ell-2}$, such that for every integral multilinear form $m$ in the coefficients of $A_1, \ldots, A_{\ell-1}$, we have

$$mF_i^{(\ell-1)} = \sum_{j=1}^{b} b_{ijm} F_j^{(\ell-1)},$$

where $b_{ijm}$ are integral linear forms in the coefficients of $A_1 A_2, \ldots, A_{\ell-1}$. There are also non-zero integral forms $G_g$ $(1 \le g \le \gamma)$ in the coefficients of the product $A_1 \ldots A_{\ell-1}$ such that every bilinear form $\mu$ in the coefficients of $A_1 \ldots A_{\ell-1}$ and $A_\ell$ satisfies $\mu G_g = \sum_{h=1}^{\gamma} c_{gh\mu} G_h$ where $c_{gh\mu}$ are integral linear forms in the coefficients of $A_1 A_2 \ldots A_\ell$. Here we use the 1st step, and the fact that if $A_1, \ldots, A_{\ell-1}$ have indeterminate coefficients then $A_1 A_2 \ldots A_{\ell-1} \neq 0$. For $F_j^{(\ell)}$ we take all products $F_i^{(\ell-1)} G_g$

$(i \leq \kappa,\ g \leq \gamma)$. Every integral multilinear form in the coefficients of $A_i$ can be written as $\sum_{t=0}^{|A_\ell|} m_t a_t^{(\ell)}$ where $m_t$ is an integral multilinear form in the coefficients of $A_1, \ldots, A_{\ell-1}$. It is sufficient to prove formula (1) for one term of the latter sum as $M$.

Now

$$E = ma_t^{(\ell)} F_i^{(\ell-1)} G_g = a_t^{(\ell)} G_g \sum_{j=1}^{\kappa} b_{ijm} F_j^{(\ell-1)} = \sum_{j=1}^{\kappa} b_{ijm} a_t^{(\ell)} G_g F_j^{(\ell-1)}.$$

Now each $b_{ijm} a_t^{(\ell)}$ is a bilinear form $\mu_j$ in coefficients of $A_1 \cdot \ldots \cdot A_{\ell-1}$ and $A_\ell$ and therefore

$$E = \sum_{j=1}^{\kappa} \sum_{h=1}^{\gamma} c_{gh\mu_j} G_h F_j^{(\ell-1)},$$

as required. Unless $A_1 \ldots A_{\ell-1} = 0$ not all $G_g$ vanish after specialization. Therefore the vanishing of all $F_i^{(\ell-1)} G_g$ $(i \leq \kappa,\ g \leq \gamma)$ implies the vanishing of all $F_i^{(\ell-1)}$, but by the inductive assumption this is possible only if $A_1 \ldots A_{\ell-2} = 0$.

<u>3rd step.</u> $\mathbf{x} = \langle x_1, \ldots, x_n \rangle$. We make Kronecker's substitution and obtain $S_d B = S_d A_1 \ldots S_d A_\ell$.

If $A_i$ are complete polynomials in $\mathbf{x}$ with $|A_i| = d-1$ and indeterminate coefficients then $S_d A_i$ are complete polynomials of degree $d^n-1$ with the same indeterminate coefficients, but in one variable, thus step 2 applies. If some polynomials $A_i$ satisfy $|A_i| < d-1$ we specialize coefficients suitably and profit by the fact that $S_d A_1 \ldots S_d A_{\ell-1} \neq 0$ unless $A_1 \ldots A_{\ell-1} = 0$. The proof is complete.

Theorem 10 (Kronecker 1883). *Let* $B(x) = A_1(x) \ldots A_l(x)$, *where* $A_i(x)$ *is a complete polynomial in* $x$ *with given* $|A_i|$ *and indeterminate coefficients. Every integral multilinear form in the coefficients of* $A_1, \ldots, A_l$ *satisfies an equation*

$$(3) \qquad M^{\tau} - f_1(b_0, b_1, \ldots)M^{\tau-1} + f_2(b_0, b_1, \ldots)M^{\tau-2} + \ldots = 0,$$

*where* $f_\nu$ *is an integral form of degree* $\nu$ *in the coefficients* $b_i$ *of* $B$.

Proof. Since the forms $F_i$ occurring in Theorem 9 are non-zero the determinant of the linear system of equations (1) must vanish, i.e.

$$|b_{ijM} - MI| = 0,$$

which gives the theorem.

Corollary 1. If $A(x_1, \ldots, x_n) | B(x_1, \ldots, x_n)$ and $a_0, b_0$ are the leading coefficients of $A, B$ in the antilexicographic order then for every coefficient $a_i$ of $A$ there is an integral polynomial $\Omega$ monic with respect to the first variable such that

$$\Omega\left(\frac{a_i}{a_0}, \frac{b_1}{b_0}, \frac{b_2}{b_0}, \ldots, \frac{b_r}{b_0}\right) = 0.$$

Proof. Let $B = A \cdot C$. We apply Theorem 10 to the factorization

$$\frac{B}{b_0} = \frac{A}{a_0} \cdot \frac{C}{c_0}$$

specializing the coefficients and taking

$$M_i = \frac{a_i}{a_o} \cdot \frac{c_o}{c_o} = \frac{a_i}{a_o} .$$

Corollary 2. If $A_i \in \widehat{k(\lambda)}[x]$, but $B = A_1 \ldots A_\ell$ has coefficients in $k[\lambda]$, and $|B|_\lambda \leq \gamma$, then every multilinear form in the coefficients of $A_1, \ldots, A_\ell$ that lies in $k(\lambda)$ is a polynomial in $\lambda$ of degree $\leq \gamma$.

Proof. The first assertion is an immediate consequence of (3). The second assertion follows from (3), because

$|M|_\lambda = d$ implies for some $\nu > 0$ that $d\tau \leq \nu\gamma + d(\tau - \nu)$ ,

hence $\nu d \leq \nu\gamma$ and $d \leq \gamma$.

## Section 8. Connection between reducibility and decomposability

Theorem 11 (Schinzel 1963b). *Let* $\Phi(y_1,\ldots,y_n)$ *be a polynomial over a field* $k$ *of positive degree in each* $y_i$ *and let* $F_1(\mathbf{x}_1),\ldots,F_n(\mathbf{x}_n)$ *be polynomials over* $k$, *at least two of them nonconstant. Then* $\Phi(F_1(\mathbf{x}_1),F_2(\mathbf{x}_2),\ldots,F_n(\mathbf{x}_n))$ *is reducible over* $k$ *if and only if* $F_i(\mathbf{x}_i) = G_i(H_i(\mathbf{x}_i))$, *where* $G_i \in k[u]$, $H_i \in k[\mathbf{x}_i]$ $(1 \le i \le n)$ *and* $\Phi(G_1(x_1),\ldots,G_n(x_n))$ *is reducible over* $k$.

We remind the reader that for $i \neq j$ the sequences $\mathbf{x}_i, \mathbf{x}_j$ contain distinct variables.

Lemma. Let $\Psi(z,y_1,\ldots,y_\ell)$ and $F(x)$ be polynomials over $k$, the former of positive degree in each variable, the latter non-constant. If

$\Psi(F(\mathbf{x}),y_1,\ldots,y_\ell)$ is reducible over $k$, then $F(\mathbf{x}) = G(H(\mathbf{x}))$,

$G \in k[u]$, $H \in k[\mathbf{x}]$ and $\Psi(G(u),y_1,\ldots,y_\ell)$ is reducible over $k$.

Proof. Let $\mathbf{y} = \langle y_1,\ldots,y_\ell\rangle$ and suppose

$$\Psi(z,\mathbf{y}) = \sum_{j=0}^{m} A_j(z)M_j(\mathbf{y}),$$

where $M_j$ are monomials ordered antilexicographically and $A_0(z) \neq 0$. When $d(z) = \underset{0\le j\le m}{\text{g.c.d.}} A_j(z) \neq 1$ the polynomial $\Psi(F(\mathbf{x}),\mathbf{y})$ is trivially reducible since

$$d(F(\mathbf{x})) \mid \Psi(F(\mathbf{x}),\mathbf{y}).$$

and the lemma holds with $G = u$, $H = F$.

Now assume $\underset{0\le j\le m}{\text{g.c.d.}}\, A_j = 1$. Then there exist polynomials $D_j(z)$ with $\sum_{j=0}^{m} A_j(z)D_j(z) = 1$. The substitution $z = F(\mathbf{x})$ gives $\underset{0\le j\le m}{\text{g.c.d.}}\, A_j(F(\mathbf{x})) = 1$. Let $\Psi(F(\mathbf{x}),\mathbf{y})$ be reducible, say

$$\Psi(F(\mathbf{x}),\mathbf{y}) = P(\mathbf{x},\mathbf{y})Q(\mathbf{x},\mathbf{y}), \quad \text{where}$$

$$P(\mathbf{x},\mathbf{y}) = \sum_{j=0}^{p} B_j(\mathbf{x})P_j(\mathbf{y}) \in k[\mathbf{x},\mathbf{y}]\setminus k, \quad B_0(\mathbf{x}) \ne 0$$

$$Q(\mathbf{x},\mathbf{y}) = \sum_{j=0}^{q} C_j(\mathbf{x})Q_j(\mathbf{y}) \in k[\mathbf{x},\mathbf{y}]\setminus k, \quad C_0(\mathbf{x}) \ne 0$$

and $P_j, Q_j$ are monomials arranged antilexicographically. We have $\deg_{\mathbf{y}} P > 0$, $\deg_{\mathbf{y}} Q > 0$ since otherwise $P \mid \underset{0\le j\le m}{\text{g.c.d.}} A_j(F)$ and $P \in k$ or the same holds for $Q$. By Corollary 1 to Theorem 10 there exist polynomials $\Omega_j, \Omega_j'$ over $k$ monic with respect to the first variable, such that

$$\Omega_j\left(\frac{B_j(\mathbf{x})}{B_0(\mathbf{x})}, \frac{A_1(F(\mathbf{x}))}{A_0(F(\mathbf{x}))}, \ldots, \frac{A_m(F(\mathbf{x}))}{A_0(F(\mathbf{x}))}\right) = 0,$$

$$\Omega_j'\left(\frac{C_j(\mathbf{x})}{C_0(\mathbf{x})}, \frac{A_1(F(\mathbf{x}))}{A_0(F(\mathbf{x}))}, \ldots, \frac{A_m(F(\mathbf{x}))}{A_0(F(\mathbf{x}))}\right) = 0.$$

It follows that

$$K = k\left(\frac{B_1(\mathbf{x})}{B_0(\mathbf{x})}, \ldots, \frac{B_p(\mathbf{x})}{B_0(\mathbf{x})}, \frac{C_1(\mathbf{x})}{C_0(\mathbf{x})}, \ldots, \frac{C_q(\mathbf{x})}{C_0(\mathbf{x})}, \quad F(\mathbf{x})\right)$$

is algebraic over $k(F(\mathbf{x}))$. Since $\text{tr.deg.}k(F(\mathbf{x}))/k = 1$ we have $\text{tr.deg.}K/k = 1$. Theorem 4 implies the existence of an $H \in k[\mathbf{x}]$ such that $\frac{B_j(\mathbf{x})}{B_o(\mathbf{x})} = b_j(H(\mathbf{x}))$, $\frac{C_1(\mathbf{x})}{C_o(\mathbf{x})} = c_j(H(\mathbf{x}))$,

$F(\mathbf{x}) = G(H(\mathbf{x}))$, $b_j, c_j \in k(u)$ $G \in k[u]$. Since $B_o(\mathbf{x})C_o(\mathbf{x}) = A_o(F(\mathbf{x}))$ we get

$$\frac{\Psi(G(H(\mathbf{x})),\mathbf{y})}{A_o(G(H(\mathbf{x})))} = \sum_{j=0}^{p} b_j(H(\mathbf{x}))P_j(\mathbf{y}) \sum_{j=0}^{q} c_j(H(\mathbf{x}))Q_j(\mathbf{y}).$$

Replacing $H(\mathbf{x})$ by $u$, we get

$$\frac{\Psi(G(u),\mathbf{y})}{A_o(G(u))} = \sum_{j=0}^{p} b_j(u)P_j(\mathbf{y}) \sum_{j=0}^{q} c_j(u)Q_j(\mathbf{y})$$

and by Gauss's lemma $\Psi(G(u),\mathbf{y})$ is reducible in $k[u,\mathbf{y}]$ (see van der Waerden 1970 § 5.4).

Proof of Theorem 11. The condition is clearly sufficient. In order to prove its necessity we proceed by induction on the total number of variables. The theorem is trivial for the number of variables equal to $n$. Suppose the theorem is true for less than $N$ variables, $N > n$. Suppose $\mathbf{x}_1$ consists of $m > 1$ variables. Label other variables occurring $y_1,\ldots,y_\ell$. Set $\Psi(z,y_1,\ldots,y_\ell) = \Phi(z,F_2(\mathbf{x}_2),\ldots,F_n(\mathbf{x}_n))$, so that $\Psi(F(\mathbf{x}_1),y_1,\ldots,y_\ell) = \Phi(F(\mathbf{x}_1),F(\mathbf{x}_2),\ldots,F(\mathbf{x}_n))$ and assuming reducibility of the latter polynomial apply the lemma. We get $F_1 = G_1(H_1(\mathbf{x}))$ and

$\Psi(G_1(u),y_1,\ldots,y\ )$ is reducible over $k$, i.e.

$\Phi(G_1(u),F_2(\mathbf{x}_2),\ldots,F_n(\mathbf{x}_n))$ is reducible.

The total number of variables now is $N-m+1 < N$. The inductive assumption applies and the theorem follows.

Definition 9. *A polynomial* $L(x)$ *is additive if* $L(x+y) = L(x) + L(y)$.

Corollary 1. If char $k = 0$, all additive polynomials are of the form $ax$; if char $k = p$, they are of the form $\sum_{i=0}^{j} a_i x^{p^i}$.

Proof. If $F(x) = \Sigma a_\ell x^\ell$ is additive all terms must be additive. Thus $a_\ell \neq 0$ implies $(x+y)^\ell = x^\ell + y^\ell$ hence

$$\begin{cases} \ell = 1 & \text{if char } k = 0 \\ \ell = p^i,\ i \geq 0 & \text{if char } k = p. \end{cases}$$

Theorem 12. *Let* $F, G, H \in k[x]$ *and* $|F||G||H| \neq 0$. *Then* $F(x) + G(y) + H(z)$, *is reducible over* $k$ *if and only if* $F(x) - F(0) = L(F_1(x))$, $G(y) - G(0) = L(G_1(y))$, $H(z) - H(0) = L(H_1(z))$, *where* $L$ *is additive and* $L(t) + F(0) + G(0) + H(0)$ *is reducible over* $k$.

Lemma 1. If

$$(1) \quad F(x) + G(y) = H_1(H_2(x,y)),$$

where $F(0) + G(0) = 0 = H_2(0,0)$, then $H_1$ is additive and $H_2(x,y) = H_2(x,0) + H_2(0,y)$.

Proof. Substituting $y = 0$ in (1) we get

$$(2) \quad F(x) + G(0) = H_1(H_2(x,0)).$$

Subtracting (2) from (1) gives

$G(y) - G(0) = H_1(H_2(x,y)) - H_1(H_2(x,0)) \equiv 0 \mod H_2(x,y) - H_2(x,0)$,

hence $H_2(x,y) = H_2(x,0) + R(y)$. Setting $x = 0$ gives $H_2(x,y) = H_2(x,0) + H_2(0,y)$. Let

$H_1 = \Sigma\, a_\ell x^\ell$. It follows that

$a_\ell(H_2(x,0) + H_2(0,y))^\ell$ must be the sum of a polynomial in

$x$ and a polynomial in $y$. Hence either $a_\ell = 0$ or $\ell = 0$

or $\ell = p^i$, char $k = p$; whence $H_1(t) = L(t) + c$ where

$L$ is additive, $F(x) + G(y) = L(H_2(x,y)) + c$. Letting

$x = y = 0$ gives $c = 0$.

Lemma 2. If $F(x) + G(y) + H(z)$ is reducible over $k$, then

$$F(x) - F(0) = L(F_1(x)),$$

$$G(y) - G(0) = L(G_1(y)),$$

$L$ is additive and $F(0) + G(0) + L(t) + H(z)$ is reducible over $k$.

Proof. $F(x) + G(y) + H(z) = \Psi(F(x) + G(y) - F(0) - G(0), H(z))$,

where $\Psi(u,v) = u + v + F(0) + G(0)$. So by the Lemma to Theorem 11

$$F(x) + G(y) - F(0) - G(0) = H_1(H_2(x,y)) \text{ and } H_1(t) + H(z) + $$

$$+ F(0) + G(0) \text{ is reducible.}$$

Lemma 2 follows now from Lemma 1 (Without loss of generality we may assume that $H_2(0,0) = 0$.).

Proof of Theorem 12. If $F(x) + G(y) + H(z)$ is reducible

over $k$ we have conclusions of Lemma 2:

$$F(0) + G(0) + L(t) + H(z) \text{ is reducible over } k.$$

Let $t = u + v$: $F(0) + G(0) + L(u) + L(v) + H(z)$

is reducible. Apply Lemma 2. We get

$$L(v) = L_1(A(v)),$$

$$H(z) - H(0) = L_1(H_1(z)) \text{ and}$$

$$P(u,w) = F(0) + G(0) + H(0) + L(0) + L(u) + L_1(w) \text{ is}$$

reducible over $k$. However $L(0) = 0$ and $L_1$ is additive so that $P(u,w) = F(0) + G(0) + H(0) + L_1(A(u) + w)$ and the theorem follows in one direction with $L_1$ in place of $L$. The converse is obvious from the additivity of $L$.

Corollary 2. (**Ehrenfeucht** and Pełczyński, see Schinzel 1963a). If char $k = 0$, then $F(x) + G(y) + H(z)$ is irreducible over $k$.

Proof. Immediate: $L(t) + c = at + c$ is irreducible.

Corollary 3. (Tverberg 1966). If $k$ of characteristic $p$ is algebraically closed then $F(x) + G(y) + H(z)$ is reducible over $k$ if and only if it can be represented as

$$F_1(x)^p + cF_1(x) + G_1(y)^p + cG_1(y) + H_1^p(z) + cH_1(z), \ c \in k,$$

$$F_1, G_1, H_1 \in k[x].$$

Proof. The sufficiency of the condition is obvious. To prove the necessity Tverberg noted: If $L$ is additive of degree $> 1$ over an algebraically closed field of characteristic $p$ then $L = M^p + cM$. Indeed, suppose

$$L = x^{p^h} + \sum_{i=1}^{h} b_i x^{p^{h-i}} .$$

Then $c$ is a non-zero solution to $c^{1+p+\ldots+p^{h-1}} =$

$$= \sum_{i=1}^{h} (-1)^{i-1} b_i^{p^{i-1}} c^{p^i+p^{i+1}+\ldots+p^{h-1}} \quad \text{and}$$

$$M = x^{p^{h-1}} + \sum_{i=1}^{h-1} x^{p^{h-1-i}} \sum_{j=1}^{h-i} (-1)^{j+1} b_{i+1}^{p^{i-1}} c^{-(1+p+\ldots+p^{j-1})} .$$

From the theorem we get

$$F(x) - F(0) = M(F_0(x))^p + cM(F_0(x)) .$$

We can find $\gamma$ in $k$ so that

$$F(x) = (M(F_0(x) + \gamma))^p + c(M(F_0(x) + \gamma)),$$

take $F_1(x) = M(F_0(x) + \gamma)$ and similarly for $G$ and $H$.

The question when $F(x) + G(y)$ is reducible is much harder even for $k = \mathbb{C}$. Clearly

$$F_1(x) - G_1(y) \mid F(x) - G(y) \quad \text{if} \quad \begin{array}{l} F = H(F_1(x)), \\ \qquad\qquad\qquad |H| > 1 \\ G = H(G_1(y)). \end{array}$$

Davenport and Lewis found another example of reducibility, namely $F(x) = T_4(x)$, $G(y) = T_4(y)$. Here

$$F(x) + G(y) = (x^2 + \sqrt{2}\,xy + y^2 - 2)\,(x^2 - \sqrt{2}\,xy + y^2 - 2).$$

For further instances of reducibility over $\mathbb{C}$ of polynomials $F(x) + G(y)$ see M.Fried 1973.

M.Fried 1970 proved that $\frac{F(x)-F(y)}{x-y}$ is reducible in $k$ provided $k = \hat{k}$, char $k = 0$ or char $k > |F|$ if and only if either $F = G(H(x))$, $G,H \in k[x]$, $|G| > 1$, $|H| > 1$ or $F(x) = L_1 \circ x^p \circ L_2$ or $F(x) = L_1 \circ T_p \circ L_2$, $L_1, L_2 \in k[x]$, $|L_1| = |L_2| = 1$.

## Section 9. Elimination theory for systems of homogeneous equations

<u>Definition 10.</u> *The coefficient vector of a form of degree* $d$ *is the sequence of coefficients of all monomials of degree* $d$ *ordered lexicographically.*

<u>Theorem 13</u> (Mertens 1899 for $\hat{k}$ of characteristic 0, van der Waerden 1926 for all $k$). *For every system* $S$ *of* $r$ *homogeneous equations with* $n$ *variables, of given degrees* $d_1,\ldots,d_r$ *and with indeterminate coefficient vectors* $\mathbf{a}_1,\ldots,\mathbf{a}_r$ *there exist integral forms* $R_j(\mathbf{a}_1,\ldots,\mathbf{a}_r)$ $(1 \le j \le s)$ *with the following property. The system obtained from* $S$ *by a substitution* $\mathbf{a}_i = \mathbf{a}_i^*$, *where the components of* $\mathbf{a}_i^*$ *belong to a field* $k$ *has a non-zero solution in* $k^n$ *if and only if* $R_j(\mathbf{a}_1^*,\mathbf{a}_2^*,\ldots,\mathbf{a}_r^*) = 0$ *for* $j = 1,2,\ldots,s$.

<u>Proof.</u> Let the system $S$ consist of equations $F_i(\mathbf{x}) = 0$ $(1 \le i \le r)$.

Consider first $n = 1$. Then

$$F_i = a_i x^{d_i} \qquad (1 \le i \le r),$$

and we take

$$\{R_1,\ldots,R_s\} = \{a_1,\ldots,a_r\}.$$

Next, consider $n = 2$, $r = 2$,

$$F_1 = a_o x^d + \ldots + a_d y^d,$$

$$F_2 = b_o x^e + \ldots + b_e y^e.$$

The resultant

$$\begin{matrix} e \left\{ \begin{matrix} \\ \\ \end{matrix} \right. \\ d \left\{ \begin{matrix} \\ \\ \end{matrix} \right. \end{matrix} \begin{vmatrix} a_o \cdots\cdots\cdots a_d 0 \ldots 0 \\ 0 \ldots 0 \; a_o \cdots\cdots\cdots a_d \\ b_o \cdots\cdots\cdots b_e 0 \ldots 0 \\ 0 \ldots 0 b_o \cdots\cdots\cdots b_e \end{vmatrix} = \Sigma \pm a_{\mu_1} \ldots a_{\mu_e} b_{\nu_1} \ldots b_{\nu_d}$$

has the required property (see van der Waerden 1970, §5,8).

Moreover if the product

$$(1) \qquad a_{\mu_1} \ldots a_{\mu_e} b_{\nu_1} \ldots b_{\nu_d}$$

occurs in the above sum, then

$$a_{\mu_i} \text{ is in } (\mu_i + i)^{\text{th}} \text{ column,}$$

$$b_{\nu_j} \text{ is in } (\nu_j + j)^{\text{th}} \text{ column,}$$

hence $\sum_{i=1}^{e} \mu_i + \sum_{j=1}^{d} \nu_j = (d+e)\frac{d+e+1}{2} - \frac{d(d+1)}{2} - \frac{e(e+1)}{2} = de.$

If we define weight of the product (1) as the sum of indices the obtained equality can be expressed by saying that the resultant of $F_1$ and $F_2$ is isobaric of weight $de$.

Suppose for a moment that all $F_i$ are of the same degree $d$ and let $n = 2$, $r > 2$. Consider the forms

$$U(x_1, x_2) = u_1F_1 + u_2F_2 + \ldots + u_rF_r,$$

$$V(x_1, x_2) = v_1F_1 + v_2F_2 + \ldots + v_rF_r.$$

If $F_i(x_1,x_2) = 0$ $(1 \leq i \leq r)$ is solvable in $k^2 \setminus (0,0)$, $U,V$ have a common factor in $k[\mathbf{u},\mathbf{v}]$, hence

$$R(\mathbf{u},\mathbf{v}) = 0 ,$$

where $R$ is the resultant of $U,V$. Conversely, if $R(\mathbf{u},\mathbf{v})=0$ then $U,V$ have a common factor in $k[\mathbf{u},\mathbf{v}]$ which implies the existence of a common non-trivial zero of $F_i$ $(1 \leq i \leq r)$. The resultant $R$ is a polynomial in the coefficients of $U,V$:

$$R = \Sigma\, u_1^{\alpha_{i1}} \cdots u_r^{\alpha_{ir}} v_1^{\beta_{i1}} \cdots v_r^{\beta_{ir}} D_i(\mathbf{a}_1,\ldots,\mathbf{a}_r) .$$

Necessary and sufficient condition for the equality R=O to hold identically in $\mathbf{u},\mathbf{v}$ is that all $D_1(\mathbf{a}_1,\ldots,\mathbf{a}_r) = 0$. Moreover the $D_i$ are integral forms of degree $2d$ with all terms of weight $d^2$.

Now we prodeed to complete the proof of TXeorem 13 by induction.

Suppose theorem 13 established for equations in less than $n$ variables, all of the same degree. Let $F_i(x_1,\ldots,x_n)$ $(i \leq r)$ be forms of degree $d$ and let

$$G_i(t,x_n) = F_i(\mathbf{t}x_1,\ldots,tx_{n-1},x_n) = \sum_{j=0}^{d} H_{ij}(x_1,\ldots,x_{n-1})\, t^j x_n^{d-j} .$$

The system $F_i(x_1,\ldots,x_n) = 0$ $(i \leq r)$ has a non-zero solution in $\hat{k}^n$ if and only if there exist $(x_1,\ldots,x_{n-1}) \in \hat{k}^{n-1} \setminus 0$ and $(t,x_n) \in k^2 \setminus (0,0)$ such that $G_i(t,x_n) = 0$ $(i \leq r)$. By the preceding argument, necessary and sufficient condition for solvability of the system

$G_i(t,x_n) = 0 \quad (i \le r) \quad \text{in} \quad (t,x_n) \in \hat{k}^2 \setminus (0,0) \quad \text{is}$

$$(2) \qquad D_1(\ldots,H_{ij}(x_1,\ldots,x_{n-1}),\ldots) = \ldots =$$

$$= D_g(\ldots,H_{ij}(x_1,\ldots,x_{n-1}),\ldots) = 0.$$

Observe that $\deg H_{ij} = j$. Hence all polynomials (2) are forms of the same degree $d^2$ in $x_1,\ldots,x_{n-1}$. Moreover their coefficients are integral forms of degree $2d$ in the coefficients of $F_1,\ldots,F_r$. By the inductive assumption (2) is soluble in $(x_1,\ldots,x_{n-1}) \in \hat{k}^{n-1} \setminus \mathbf{0}$, if a certain system of integral forms in the coefficients of $F_1,\ldots,F_r$ vanishes. Our special case of the theorem follows by induction.

If the $F_i$ are not all of the same degree, say $F_i$ is of degree $d_i$ we consider the forms

$$G_{ij} = x_j^{d-d_i} F_i \qquad j \le n, \quad d = \max\{d_i\}.$$

An application of the above special case proves the theorem.

Remark. The above proof which is a modification of the one by Kapferer 1929 shows that forms $R_j(\mathbf{a}_1,\ldots,\mathbf{a}_r)$ can be chosen all of degree $2^{n-1} d^{2^{n-1}-1}$.

Theorem 14. (Mertens 1899). *For every system $S_0$ of $n$ homogeneous equations with $n$ variables, of given degrees $d_1,\ldots,d_n$ and with indeterminate coefficient vectors $\boldsymbol{a}_1,\ldots,\boldsymbol{a}_n$ there exists a unique integral form $R_0(\boldsymbol{a}_1,\ldots,\boldsymbol{a}_n)$ with the following properties:*

*(3) $R_0$ is irreducible over $\mathbf{Q}$ and primitive.*

(4) *The system obtained from* $S_o$ *by a substitution* $a_i = a_i^*$, *where the components of* $a_i^*$ *belong to a field* $k$, *has a non-zero solution in* $k^n$ *if and only if* $R_o(a_1^*,\ldots,a_n^*) = 0$.

(5) *If* $a_i^*$ *is the coefficient vector of the form* $x_i^{d_i}$ $(1 \leq i \leq r)$ *then* $R_o(a_1^*,\ldots,a_n^*) > 0$.

(6) $R_o(a_1,\ldots,a_{n-1},0) = 0$.

For the proof it will be convenient to introduce the following

Convention. An integral resultant for a system $S$ of $r$ homogeneous equations with $n$ variables, of given degrees $d_1,\ldots,d_r$ and with indeterminate coefficient vectors $\mathbf{a}_1,\ldots,\mathbf{a}_r$ is an integral polynomial $R(\mathbf{a}_1,\ldots,\mathbf{a}_r)$ with the following property. If the system obtained from $S$ by a substitution $\mathbf{a}_i = \mathbf{a}_i^*$, where the components of $\mathbf{a}_i^*$ belong to a field $k$, has a non-zero solution in $\hat{k}^n$ then $R(\mathbf{a}_1^*,\ldots,\mathbf{a}_r^*) = 0$.

Lemma 1. Let the system $S$ consist of equations $F_i(\mathbf{x}) = 0$ $(1 \leq i \leq r)$, let $a_{ij}$ be the coefficient of $x_j^{d_i}$ in $F_i$ and let $\mathbf{a}_i = (\mathbf{a}'_{ij}, a_{ij}, \mathbf{a}''_{ij})$, where $\mathbf{a}'_{ij}, \mathbf{a}''_{ij}$ are suitable vectors. By the definition of the coefficient vector $\mathbf{a}'_{i1} = \mathbf{a}''_{in}$ are empty. For every integral polynomial $R$ the following five conditions are equivalent.

(7) $R$ is an integral resultant for $S$,

(8) $R\left(\mathbf{a}'_{1n}, a_{1n} - \frac{F_1(\mathbf{t})}{t_n^{d_1}}, \mathbf{a}'_{2n}, a_{2n} - \frac{F_2(\mathbf{t})}{t_n^{d_2}}, \ldots, \mathbf{a}'_{rn}, a_{rn} - \frac{F_r(\mathbf{t})}{t_n^{d_r}}\right) = 0,$

(9) There exist an integer $\mu$ and integral polynomials $H_{in}(\mathbf{a}_1,\ldots,\mathbf{a}_r,\mathbf{x})$ such that

$$R(\mathbf{a}_1,\ldots,\mathbf{a}_r)x_n^{\mu} = \sum_{i=1}^{r} H_{in}(\mathbf{a}_1,\ldots,\mathbf{a}_r,\mathbf{x})F_i(\mathbf{x}),$$

(10) For every $j \le n$

$$R\left(\mathbf{a}'_{1j}, a_{1j} - \frac{F_1(\mathbf{t})}{t_j^{d_1}}, \mathbf{a}''_{1j}, \mathbf{a}'_{2j}, a_{2j} - \frac{F_2(\mathbf{t})}{t_j^{d_2}}, \mathbf{a}''_{2j}, \ldots, \mathbf{a}'_{rj}, a_{rj} - \frac{F_r(\mathbf{t})}{t_j^{d_r}}, a''_{rj}\right) = 0,$$

(11) For every $j \le n$ there exist integers $\mu_j$ and integral polynomials $H_{ij}(\mathbf{a}_1,\ldots,\mathbf{a}_r,\mathbf{x})$ such that

$$R(\mathbf{a}_1,\ldots,\mathbf{a}_r)x_j^{\mu_j} = \sum_{i=1}^{r} H_{ij}(\mathbf{a}_1,\ldots,\mathbf{a}_r,\mathbf{x})F_i(\mathbf{x}).$$

Proof. Implication (7) $\longrightarrow$ (8). The system obtained from S by substitution $\mathbf{a}_i \longrightarrow \left(\mathbf{a}'_{in}, a_{in} - \frac{F_i(\mathbf{t})}{t_n^{d_i}}\right)$ has a non-zero solution in $k = \mathbb{Q}(\mathbf{t})$, namely $\mathbf{x} = \mathbf{t}$.

Implication (8) $\longrightarrow$ (9). From the Taylor formula we have

$$R(\mathbf{a}'_{1n}, a_{1n}-u_1, \mathbf{a}'_{2n}, a_{2n}-u_2, \ldots, \mathbf{a}'_{rn}, a_{rn}-u_r) =$$

$$= R(\mathbf{a}_1,\mathbf{a}_2,\ldots,\mathbf{a}_r) - \sum_{\alpha\in A} G_\alpha(\mathbf{a}_1,\ldots,\mathbf{a}_n)u_1^{\alpha_1}u_2^{\alpha_2}\ldots u_r^{\alpha_r},$$

where $A$ is a finite set of vectors $\alpha = (\alpha_1,\ldots,\alpha_r)$ satisfying $\alpha_i \ge 0$, $\alpha_1+\alpha_2+\ldots+\alpha_r > 0$ and $G_\alpha$ are integral polynomials. Taking

$$\mu = \max_{\alpha\in A} \alpha_1 d_1 + \ldots + \alpha_r d_r$$

and substituting $u_i = \frac{F_i(\mathbf{t})}{t_n^{d_i}}$ we get (9) with $\mathbf{x}$ replaced by $\mathbf{t}$.

<u>Implication</u> (9) ⟶ (10). Let $j$ be fixed. Substituting in (9) $a_{ij} - \frac{F_i(\mathbf{t})}{t_j^{d_i}}$ for $a_{ij}$ $(1 \le i \le r)$, $\mathbf{x} = \mathbf{t}$ we get 0 on the right hand side (all terms vanish) and it remains to divide both sides by $t_n^{\mu}$. The substitution does not involve a vitious circle since $a_{ij} - \frac{F_i(t)}{t_j^{d_i}}$ does not involve $a_{ij}$.

<u>Implication</u> (10) ⟶ (11) follows in the same way as (8) ⟶ (9).

<u>Implication</u> (11) ⟶ (7). If the system obtained from S by a substitution $\mathbf{a}_i = \mathbf{a}_i^*$ has a non-zero solution $(\xi_1,\ldots,\xi_n)$ we get from (11) $R(\mathbf{a}_1^*,\ldots,\mathbf{a}_r^*)\xi_j^{\mu_j} = 0$ $(1 \le j \le n)$, hence $R(\mathbf{a}_1^*,\ldots,\mathbf{a}_r^*) = 0$.

<u>Lemma 2</u>. Integral resultants for a system S form a prime ideal in the ring of integral polynomials in the coefficients of S.

<u>Proof</u>. The only fact that needs a proof is that the ideal in question is prime. If $R_i$ $(i = 1,2)$ are integral polynomials and $R_1R_2$ is an integral resultant for S than by condition (8) of Lemma 1 we have

$$\prod_{i=1}^{2} R_i\left(\mathbf{a}'_{1n}, a_{1n} - \frac{F_1(\mathbf{t})}{t_n^{d_1}}, \mathbf{a}'_{2n}, a_{2n} - \frac{F_2(\mathbf{t})}{t_n^{d_2}}, \ldots, \mathbf{a}'_{rn}, a_{rn} - \frac{F_r(\mathbf{t})}{t_n^{d_r}}\right) = 0$$

hence the same condition is satisfied either by $R_1$ or by $R_2$.

Lemma 3. If an integral resultant $R(\mathbf{a}_1,\dots,\mathbf{a}_r)$ for a system of equations $F_i(\mathbf{x}) = 0$ $(1 \le j \le r)$ is independent of a coefficient of $F_r$ then all the coefficients of $R$ viewed as a polynomial in $\mathbf{a}_r$ are integral resultants for the system $F_i(\mathbf{x}) = 0$ $(1 \le i < r)$.

Proof. If $R$ is independent of a coefficient a of $F_r$ then also

$$E = R\left(\mathbf{a}'_{1n}, a_{1n} - \frac{F_1(\mathbf{x})}{x_n^{d_1}}, \dots, \mathbf{a}'_{r-1,n}, a_{r-1,n} - \frac{F_{r-1}(\mathbf{x})}{x_n^{d_{r-1}}}, \mathbf{a}_r\right)$$

is independent of a. But by the condition (9) of Lemma 1

$$Ex_n^{\mu} = F_r(\mathbf{x})H_r\left(\mathbf{a}'_{1n}, a_{1n} - \frac{F_1(\mathbf{x})}{x_n^{d_1}}, \dots, \mathbf{a}'_{r-1,n}, a_{r-1,n} - \frac{F_{r-1}(\mathbf{x})}{x_n^{d_{r-1}}}, \mathbf{a}_r, \mathbf{x}\right)$$

(All the remaining terms of the sum on the right hand side of (9) vanish on substitution $a_{in} - \frac{F_i(\mathbf{x})}{x_n^{d_i}}$ for $a_{in}$).

Since $F_r(\mathbf{x})$ depends on a we must have

$$E = 0. \tag{12}$$

Now, let

$$R = \Sigma R_m(\mathbf{a}_1,\dots,\mathbf{a}_{r-1})M_m(\mathbf{a}_r),$$

where $M_m$ are distinct monomials in $\mathbf{a}_r$. The equality (12) implies that for all $m$

$$R_m\left(\mathbf{a}'_{1n}, a_{1n} - \frac{F_1(\mathbf{x})}{x_n^{d_1}}, \dots, \mathbf{a}'_{r-1,n}, a_{r-1,n} - \frac{F_{r-1}(\mathbf{x})}{x_n^{d_{r-1}}}\right) = 0$$

and by the condition (9) of Lemma 1 again the polynomials $R_m$ are integral resultants for the system $F_i(\mathbf{x}) = 0$ $(1 \le i < r)$.

Lemma 4. If the number of equations in a system $S$ is less than the number $n$ of variables then the only integral resultant for $S$ is $0$.

Proof. We proceed by induction on $n$. If $n = 2$ the lemma is trivially true. Now assume its validity for less than $n$ variables, $n > 2$.

Let $R$ be an integral resultant for a system $S$ of $r$ homogeneous equations $F_i(\mathbf{x}) = 0$ $(1 \le i \le r)$ in $n$ variables, where $n > r$.

Let as before $d_i$ be the degree of $F_i, \mathbf{a}_i$ be its coefficient vector, $a_{ij}$ the coefficient of $x_j^{d_i}$ in $F_i(\mathbf{x})$ $(i \le r, j \le n)$ and $\mathbf{a}_i = (\mathbf{a}'_{ij}, a_{ij}, \mathbf{a}''_{ij})$.

Suppose that $R \ne 0$ and let $L(\mathbf{a}''_{11}, \mathbf{a}_2, \ldots, \mathbf{a}_n)$ be the leading coefficient of $R$ with respect to $a_{11}$. We do not exclude the possibility that $R$ does not depend on $a_{11}$ in which case we take $L = R$. Thus $L \ne 0$.

Put

$$L^*(\mathbf{a}''_{11}, \mathbf{a}'_{2n}, \ldots, \mathbf{a}'_{rn}, \mathbf{x}) = L\left(\mathbf{a}''_{11}, \mathbf{a}'_{2n}, a_{2n} - \frac{F_2(\mathbf{x})}{x_n^{d_2}}, \ldots, \mathbf{a}'_{rn}, a_{rn} - \frac{F_r(\mathbf{x})}{x_n^{d_r}}\right).$$

By condition (9) of Lemma 1

$$R(\mathbf{a}_1,\mathbf{a}'_{2n},a_{2n} - \frac{F_2(\mathbf{x})}{x_n^{d_2}},\ldots,\mathbf{a}'_{rn},a_{rn} - \frac{F_r(\mathbf{x})}{x_n^{d_r}})x_n^{\mu} =$$

$$= H_1(\mathbf{a}_1,\mathbf{a}'_{2n},a_{2n} - \frac{F_2(\mathbf{x})}{x_n^{d_2}},\ldots,\mathbf{a}'_{rn},a_{rn} - \frac{F_r(\mathbf{x})}{x_n^{d_r}},\mathbf{x})F_1(\mathbf{x}).$$

(All the remaining terms on the right hand side of (9) vanish on the substitution).
Comparing on both sides the leading coefficients with respect to $a_{11}$ we get

$$L^* x_n^{\mu} = G(\mathbf{a}''_{11},\mathbf{a}'_{2n},\ldots,\mathbf{a}'_{rn},\mathbf{x})x_1^{d_1},$$

where for a suitable $\nu$ : $x_n^{\nu}$ G is an integral polynomial, possibly 0. Substituting $x_1 = 0$ we get

$$(13) \qquad L^*(\mathbf{a}''_1,\mathbf{a}'_{2n},\ldots,\mathbf{a}'_{rn},0,x_2,\ldots,x_n) = 0.$$

Let $\mathbf{a}''_{11} = (\mathbf{b}_1,\mathbf{c}_1)$, $\mathbf{a}'_{in} = (\mathbf{b}_i,\mathbf{c}_i)$ $(2 \le i \le r)$, where the components of $\mathbf{b}_i$ are coefficients of the terms of $F_i$ containing $x_1$ and the components of $\mathbf{c}_i$ are the remaining coefficients of $F_i$ $(1 \le i \le r)$. We can write

$$L = \sum_p L_p(\mathbf{c}_1,\mathbf{c}_2,a_{2n},\ldots,\mathbf{c}_r,a_{rn})M_p(\mathbf{b}_1,\ldots,\mathbf{b}_r),$$

where $M_p$ are distinct monomials in $\mathbf{b}_1,\ldots,\mathbf{b}_r$. The equality (13) gives

$$\sum_p L_p(\mathbf{c}_1,c_2,a_{2n} - \frac{F_2(0,x_2,\ldots,x_n)}{x_n^{d_2}},\ldots$$

$$\ldots,\mathbf{c}_r,a_{rn} - \frac{F_r(0,x_2,\ldots,x_n)}{x_n^{d_r}})M_p(\mathbf{b}_1,\ldots,\mathbf{b}_r) = 0\,.$$

Since the components of $\mathbf{b}_i$ do not occur in $F_i(0,x_2,\ldots,x_n)$ it follows that for each $p$

$$L_p(\mathbf{c}_1,\mathbf{c}_2,a_{2n} - \frac{F_2(0,x_2,\ldots,x_n)}{x_n^{d_2}},\ldots,c_r,a_{rn} - \frac{F_r(0,x_2,\ldots,x_n)}{x_n^{d_r}}) = 0.$$

Hence by Lemma 1 $L_p(\mathbf{c}_1,\mathbf{c}_2,a_{2n},\ldots,\mathbf{c}_r,a_{rn})$ is for each $p$ an integral resultant for the system $F_i(0,x_2,\ldots,x_n) = 0$ $(1 \leq i \leq r)$ and by the inductive assumption $L_p(\mathbf{c}_1,\mathbf{c}_2,a_{2n},\ldots,\mathbf{c}_r,a_{rn}) = 0$. Thus $L = 0$, a contradiction.

Proof of the theorem. If 0 where the only integral resultant form (i.e. homogeneous integral resultant) — for the system $S_0$ then by Theorem 13 every system obtained from $S_0$ by the specialization of its coefficients would have a non-zero solution. But the system $x_i^{d_i} = 0$ $(1 \leq i \leq n)$ has no such solution, thus there exist non-zero integral resultant forms. Consider such forms of the least degree and among them one say $R_1$ with the least content and hence primitive. It is were reducible in $\mathbf{Z}$ one of its factors would be an integral resultant for $S_0$ by Lemma 2. Since every factor of a form is again a form this would contradict the choice of $R_1$. Thus $R_1$ satisfies the condition (3).

We assert that $R_1$ divides all integral resultants for the system $S_0$. Let $S_0$ consist of the equations $F_i(\mathbf{x}) = 0$

$(1 \le i \le n)$ and let a be a coefficient of $F_n$. If R is an integral resultant for $S_o$ then taking the resultant $R_2$ of $R_1$ and R with respect to a we get an identity

$$AR_1 + BR = R_2,$$

where $A, B, R_2$ are integral polynomials in $\mathbf{a}_1, \ldots, \mathbf{a}_n$ and $R_2$ does not depend on a. By Lemma 2 $R_2$ is an integral resultant form for $S_o$. By Lemma 3 all the coefficients of $R_2$ viewed as a polynomial in $\mathbf{a}_n$ are integral resultant forms for the system $F_i = 0$ $(1 \le i < n)$. By Lemma 4 these coefficients are 0, hence $R_2 = 0$. This means that $R_1$ and R have a common divisor and since $R_1$ is irreducible and primitive, $R/R_1$ is an integral polynomial. Thus $R_1(\mathbf{a}_1^*, \ldots, \mathbf{a}_n^*) = 0$ implies $R(\mathbf{a}_1^*, \ldots, \mathbf{a}_n^*) = 0$ for $\mathbf{a}_i^*$ in a field of arbitrary characteristic and hence by Theorem 13 $R_1$ satisfies the condition (4).

The two conditions (3) and (4) determine $R_1$ up to a sign. If $\mathbf{a}_i^*$ is the coefficient vector of the form $x_i^{d_i}$ $(1 \le i \le n)$ then by (4) $R_1(\mathbf{a}_1^*, \ldots, \mathbf{a}_n^*) \neq 0$. Therefore, the conditions (3), (4) and (5) determine $R_o$ uniquely. Lemma 4 implies (6) and the proof is complete.

<u>Definition 11</u>. *If the system* $S_0$ *consists of equations* $F_i(x) = 0$ $(1 \le i \le n)$ *the form* $R_0(\mathbf{a}_1, \ldots, \mathbf{a}_n)$ *of Theorem 14 is called the resultant of* $F_1, \ldots, F_n$ *with respect to* $x_1, \ldots, x_n$.

*Moreover* $R_0(\mathbf{a}_1^*, \ldots, \mathbf{a}_n^*)$ *is called the resultant with respect to* $x_1, \ldots, x_n$ *of the forms obtained from* $F_i$ *by substitution* $\mathbf{a}_i = \mathbf{a}_i^*$ $(1 \le i \le n)$.

The proof of Theorem 14 given above is based on van der Waerden 1928, but the proof of the crucial Lemma 4 follows the original paper of Mertens 1899. Another proof was given by Kapferer 1929.

A more circumstantial account of the properties of resultant can be found in Hurwitz 1912, Macaulay 1916 or Perron 1951. In particular, Macaulay gives an explicit expression of the resultant as the quotient of two determinants. The account given in König 1903 contains a gap on p. 265 in the proof of our Lemma 4.

Theorem 15 (E.Noether 1922). *For any polynomial $F(x)$ with given number of variables, of given degree $d$ and with an indeterminate coefficient vector $\mathbf{a}$ there exist finitely many integral forms $N_{ij}(\mathbf{a})$ $(1 \le i \le d,\ \ 1 \le j \le s_i)$ with the following property. For every algebraically closed field $k$ the polynomial $F^*$ obtained from $F$ by substitution $\mathbf{a} = \mathbf{a}^*$, where the components of $\mathbf{a}^*$ belong to $k$ and $\deg F^* = d$ is reducible over $k$ if and only if $N_{ij}(\mathbf{a}^*) = 0$ for at least one $i < d$ and all $j \le s_i$.*

Proof. (Fischer 1925). Let us consider the factorization

$$(1) \qquad aF(x) = G(x)H(x),$$

where

$$F(x) = \sum_{i_1+i_2+\ldots+i_n \le d} a_{\mathbf{i}} x_1^{i_1} x_2^{i_2} \ldots x_n^{i_n},$$

$$G(x) = \sum_{i_1+i_2+\ldots+i_n \le \beta} b_{\mathbf{i}} x_1^{i_1} x_2^{i_2} \ldots x_n^{i_n},$$

$$H(x) = \sum_{i_1+i_2+\ldots+i_n \le d-\beta} c_{\mathbf{i}} x_1^{i_1} x_2^{i_2} \ldots x_n^{i_n}.$$

Suppose $G$ fixed. The equation (1) gives a system of linear equations for $a$ and the $c_{\mathbf{i}}$. The system has a non-zero solution if and only if some determinants are 0. Since $a = 0$ implies $c_{\mathbf{i}} = 0$ the vanishing of these determinants is a necessary and sufficient condition for the existence of $H$, $a \ne 0$ satisying (1). The determinants in question are integral forms homogeneous in the $b_{\mathbf{i}}$ and of degree at most one in the $a_{\mathbf{i}}$. By Theorem 13 there exists a system of integral forms in the $\mathbf{a}_i$ the vanishing of which after the substitution $\mathbf{a} = \mathbf{a}^*$ is a necessary and sufficient condition for the existence in $k$ of non-zero $a, b_i, c_i$ satisfying (1). For each choice of $\beta = \deg G$ from the set $\{1,2,\ldots,d-1\}$ we get such a system and if $\deg F^* = d$ reducibility of $F^*$ over $\hat{k}$ is **equivalent to** the existence of $\beta$ for which all forms of the system vanish for $\mathbf{a} = \mathbf{a}^*$.

Corollary 1. (Ostrowski 1919). If a polynomial over $\mathbb{Q}$ is irreducible in $\hat{\mathbb{Q}}$, then it is irreducible in $\hat{\mathbb{F}}_p$ for all but finitely many primes $p$.

Proof. (E.Noether 1922). If $\mathbf{a}^*$ is an integer vector and $N_{ij}(\mathbf{a}^*)\neq 0$ in $\mathbb{Q}$ then $N_{ij}(\mathbf{a}^*)\neq 0$ in $\mathbb{F}_p$ for all but finitely many primes $p$.

Corollary 2. (Krull 1937). If $k_o$ is an infinite field and $F(\mathbf{x},\boldsymbol{\lambda}) \in k_o[\mathbf{x},\boldsymbol{\lambda}]$ is irreducible over $\widehat{k_o(\boldsymbol{\lambda})}$ then there exists a $\lambda^*$ with components in $k_o$ such that $F(\mathbf{x},\boldsymbol{\lambda}^*)$ is irreducible in $\hat{k}_o$.

Proof. Applying Theorem 15 with $k = k_o(\lambda)$, $d = \deg_x F(\mathbf{x},\lambda)$ $\mathbf{a}^* = \mathbf{a}(\lambda)$ we infer from irreducibility of $F(\mathbf{x},\lambda)$ over $k_o(\lambda)$ that for each $i < d$ there exists a $j_i \le s_i$ such that $N_{ij_i}(\mathbf{a}(\lambda)\nu) \neq 0$. Let $a_1(\boldsymbol{\lambda})$ be the leading coefficient of $F(\mathbf{x},\boldsymbol{\lambda})$ with respect to $x$. Since $k_o$ is infinite there exists a $\boldsymbol{\lambda}^*$ with components in $k_o$ such that $a_1(\boldsymbol{\lambda}^*) \prod_{i=1}^{d-1} N_{ij_i}(\mathbf{a}(\boldsymbol{\lambda}^*)) \neq 0$. Now by Theorem 15 applied to $k = \hat{k}_o$, $a^* = a(\lambda^*)$ the polynomial $F(x,\boldsymbol{\lambda}^*)$ is irreducible in $k_o$.

Corollary 3. If $\boldsymbol{\lambda} = (\lambda)$ ($\boldsymbol{\lambda}$ is one dimensional) then under the assumptions of Corollary 2 there exist only finitely many values of $\lambda$ such that $F(\mathbf{x},\lambda)$ is reducible in $k$.

Proof. If $N_{ij}(\mathbf{a}(\lambda)) \neq 0$ then $N_{ij}(\mathbf{a}(\lambda^*)) \neq 0$ for all but finitely many $\lambda^* \in k$.

Theorem 16. *Let* $M \in k[\mathbf{t}, t]$ *be squarefree with respect to* $t$ *over* $\hat{k(\mathbf{t})}$, $F \in k[v, \mathbf{t}, t]$ *have the leading coefficient with respect to* $v$ *prime to* $M$ *as a polynomial in* $t$ *over* $k(\mathbf{t})$. *There exists a non-zero polynomial* $\Phi \in k[v, \mathbf{t}, t]$ *such that if*

$$F(x(t), t^*, t) \equiv 0 \bmod M(t^*, t), \quad \text{for some} \quad x \in k[t], |x| < |M|_t, \ t^* \in k^r$$

*then*

$$\Phi(x(t), t^*, t) = 0$$

Proof. Let $|M|_t = m$, the leading coefficient of $M$ with respect to $t$ be $\mu(\mathbf{t})$, $|F|_v = f$, the leading coefficient of $F$ with respect to $v$ be $a(\mathbf{t}, t)$.

If $m = 0$ the condition $|x| < |M|_t$ implies $x = 0$ hence we can take $\Phi = v$.

If $m > 0$ let for indeterminates $x_0, \ldots, x_{m-1}$

$$F\left(\sum_{i=0}^{m-1} x_i t^i, \mathbf{t}, t\right) = \sum_{j=0}^{h} A_j(x_0, \ldots, x_{m-1}, \mathbf{t}) t^j \quad (h \geq m-1)$$

and let $B_j(x_0, \ldots, x_{m-1}, \mathbf{t})$ be the sum of all terms of $A_j$ of degree $f$ with respect to $x_0, \ldots, x_{m-1}$, possibly $B_j = 0$. Clearly

$$a(\mathbf{t}, t)\left(\sum_{i=0}^{m-1} x_i t^i\right)^f = \sum_{j=0}^{h} B_j(x_0, \ldots, x_{m-1}, \mathbf{t}) t^j.$$

We have for each $j \leq h$

$$\mu(\mathbf{t})^h t^j \equiv \sum_{i=0}^{m-1} \alpha_{ij} t^i \bmod M(\mathbf{t},t), \qquad \alpha_{ij} \in k[\mathbf{t}].$$

Hence

$$(2) \quad \mu(\mathbf{t})^h F(\sum_{i=0}^{m-1} x_i t^i, \mathbf{t}, t) \equiv \sum_{j=0}^{h} A_j \sum_{i=0}^{m-1} \alpha_{ij} t^i \equiv$$

$$\equiv \sum_{i=0}^{m-1} t^i \sum_{j=0}^{h} \alpha_{ij} A_j \bmod M(\mathbf{t},t)$$

and similarly

$$(3) \quad \mu(\mathbf{t})^h a(\mathbf{t},t)(\sum_{i=0}^{m-1} x_i t^i)^f \equiv \sum_{i=0}^{m-1} t^i \sum_{j=0}^{h} \alpha_{ij} B_j \bmod M(\mathbf{t},t).$$

Let us consider the system of polynomials

$$F_i(x_0,\dots,x_m,\mathbf{t},u,v) = x_m^f \sum_{j=0}^{h} \alpha_{ij} A_j (\frac{x_0}{x_m},\dots,\frac{x_{m-1}}{x_m},\mathbf{t})$$

$$(i = 0,1,\dots,m-1)$$

(4)

$$F_m(x_0,\dots,x_m,\mathbf{t},u,v) = \sum_{i=0}^{m-1} x_i u^i - x_m v,$$

where $u$ is a new indeterminate ($v$ has already appeared as argument of $F$ and $\Phi$). We assert that the resultant $R$ of $F_0,\dots,F_m$ with respect to $x_0,\dots,x_m$ is non-zero. Indeed in the opposite case for all values $u^*, v^* \in \widehat{k(\mathbf{t})}$ there would exist a non-zero solution of the system of equations $F_i(x_0,\dots,x_m,\mathbf{t},u,v) = 0$ $(0 \le i \le m)$ in $\widehat{k(\mathbf{t})}$. Let us choose $u^* \in k(\mathbf{t})$ so that

$$(5) \qquad M(\mathbf{t},u^*) = 0.$$

Then $a(\mathbf{t},u^*) \neq 0$ since $(M,a) \in k[\mathbf{t}]$ and thus $F(v,\mathbf{t},u^*) \neq 0$. Choose now $v^* \in k(\mathbf{t})$ so that

$$(6) \qquad F(v^*,\mathbf{t},u^*) \neq 0.$$

Let $\langle \xi_o,\ldots,\xi_m \rangle \in k(\mathbf{t})^{m+1}$ be a vector satisfying

$$(7) \quad F_i(\xi_o,\ldots,\xi_m,t,u^*,v^*) = 0 \qquad (0 \leq i \leq m).$$

If $\xi_m \neq 0$ then we find from (4)

$$(8) \quad \sum_{j=0}^{h} \alpha_{ij} A_j\left(\frac{\xi_o}{\xi_m},\ldots,\frac{\xi_{m-1}}{\xi_m},t\right) = 0 \qquad (i = 0,1,\ldots,m-1),$$

$$\sum_{i=0}^{m-1} \frac{\xi_i}{\xi_m} u^{*i} = v^*.$$

By (2) and (8) we get

$$\mu^h(\mathbf{t}) F\left(\sum_{i=0}^{m-1} \frac{\xi_i}{\xi_m} t^i, \mathbf{t}, t\right) \equiv 0 \bmod M(\mathbf{t},t).$$

Substituting $t = u^*$ we obtain from the two above formulae and (5)

$$F(v^*,\mathbf{t},u^*) = 0$$

contrary to (6). The contradiction shows that $\xi_m = 0$. Now (4) and (7) give

$$\sum_{j=0}^{h} \alpha_{ij} B_j(\xi_o,\ldots,\xi_{m-1},\mathbf{t}) = 0 \qquad (i = 0,1,\ldots,m-1)$$

and by (3)

$$\mu(\mathbf{t})^h a(\mathbf{t},t)\left(\sum_{i=0}^{m-1} \xi_i t^i\right)^f \equiv 0 \bmod M(\mathbf{t},t).$$

Since $M$ is squarefree with respect to $t$ over $\widehat{k(\mathbf{t})}$ and $(a,M) \in k[\mathbf{t}]$ we get $\sum_{i=0}^{m-1} \xi_i t^i \equiv 0 \bmod M(\mathbf{t},t)$ in the ring $\widehat{k(\mathbf{t})}[t]$ and since $M$ is of degree $m$ it follows that $\xi_i = 0$ $(i = 0,\ldots,m-1)$. Therefore the system $F_i(x_0,\ldots,x_m,\mathbf{t},u^*,v^*) = 0$ $(0 \le i \le m)$ has only the trivial solution and by Theorem 14 the resultant $R$ is non-zero. We set

$$\text{(9)} \qquad \Phi(v,\mathbf{t},u) = \mu(\mathbf{t})R.$$

Assume now that for a $\mathbf{t}^* \in \hat{k}^r$ and $x \in \hat{k}[t]$ we have

$$x(t) = \sum_{i=0}^{m-1} \xi_i t^i$$

and

$$F(x(t),\mathbf{t}^*,t) \equiv 0 \quad \bmod M(\mathbf{t}^*,t).$$

Then either $\mu(\mathbf{t}^*) = 0$ and by (9) $\Phi(x(t),\mathbf{t}^*,t) = 0$ or $\mu(\mathbf{t}^*) \neq 0$ and then (2) implies

$$\sum_{j=0}^{h} \alpha_{ij}(\mathbf{t}^*)A_j(\xi_0,\ldots,\xi_{m-1},\mathbf{t}^*) = 0 \qquad (0 \le i \le m-1).$$

This gives by (4)

$$F_i(\xi_0,\ldots,\xi_{m-1},1,\mathbf{t}^*,t,x(t)) = 0 \qquad (0 \le i \le m)$$

Thus by Theorem 14 $R = 0$ and by (6)

$$\Phi(x(t),\mathbf{t}^*,t) = 0$$

Definition 12. *A polynomial in several variables is monic if the coefficient of the leading term in the antilexicographic order equals 1.*

Theorem 17. (Salomon 1915). *Let $k$ be an algebraically closed field and $k(\lambda)^{sep}$ the separable closure of $k(\lambda)$. If $F(x,\lambda)$ is irreducible over $k(\lambda)$ then all monic factors of $F(x,\lambda)$ irreducible over $k(\lambda)^{sep}$ are conjugate over $k(\lambda)$ and the dimension of the linear space over $k$ spanned by the coefficients of any one factor does not exceed $\deg_\lambda F$.*

Proof. (following Krull 1937). Let $I(\mathbf{x})$ be a monic factor of $F(\mathbf{x},\lambda)$ irreducible over $k(\lambda)$ and let $1,\rho_1,\dots,\rho_\mu$ be a basis for the linear space spanned over $k$ by the coefficients of $I(\mathbf{x})$. We have

$$I(\mathbf{x}) = \phi_0(\mathbf{x}) + \rho_1\phi_1(\mathbf{x}) + \dots + \rho_\mu\phi_\mu(\mathbf{x}),\quad \phi_i(x) \in k[x]\,,$$

and $K = k(\lambda,\rho_1,\dots,\rho_\mu)$ is of finite degree $n$ over $k(\lambda)$. Denoting by superscript $(\nu)$ the conjugates over $k(\lambda)$ we have

$$I^{(\nu)}(\mathbf{x}) = \phi_0(\mathbf{x}) + \rho_1^{(\nu)}\phi_1(\mathbf{x}) + \dots + \rho^{(\nu)}\phi_\mu(\mathbf{x}) \qquad (1 \le \nu \le n)$$

different for different $\nu$'s by separability. Since $I^{(\nu)}(\mathbf{x})$ are monic and irreducible in $k(\lambda)^{sep}$ they are also relatively prime.

We have

$$I^{(\nu)}(\mathbf{x}) \,\big|\, F(\mathbf{x},\lambda), \quad \text{so}$$

$$\prod_{\nu=1}^{n} I^{(\nu)}(\mathbf{x}))\,|\,F(\mathbf{x},\lambda).$$

But the product of all the conjugates, $\prod_{\nu=1}^{n} I^{(\nu)}(\mathbf{x}) \in k(\lambda)[\mathbf{x}]$, hence by the irreducibility of $F(\mathbf{x},\lambda)$

$$F(\mathbf{x},\lambda) = h(\lambda) \prod_{\nu=1}^{n} [\phi_o(\mathbf{x}) + \rho_1^{(\nu)}\phi_1(\mathbf{x}) + \ldots + \rho_\mu^{(\nu)}\phi_\mu(\mathbf{x})].$$

This proves the first part of the theorem. In order to prove the second part consider the form

$$H(u_o,\ldots,u_\mu) = h(\lambda) \prod_{\nu=1}^{n} (u_o + \rho_1^{(\nu)} u_1 + \ldots + \rho_\mu^{(\nu)} u_\mu) \in k(\lambda)[u_o,\ldots,u_\mu].$$

The coefficients of $H$ are multilinear forms in the $\rho_j^{(\nu)}$'s, i.e. in the coefficients of $I^{(\nu)}(\mathbf{x})$.
By Corollary 2 to Theorem 10, the coefficients of monomials in $u_o,\ldots,u_\mu$ are in $k[\lambda]$ and are of degree $\leq \gamma = \deg_\lambda F$.

On the other hand, the above form $H$ is

$$\Psi_o(\mathbf{u})\lambda^\gamma + \Psi_1(\mathbf{u})\lambda^{\gamma-1} + \ldots + \Psi_\gamma(\mathbf{u}),$$

where the $\Psi_i$ are forms in $u_o,\ldots,u_\mu$.
If $\mu > \gamma$, the system of equations $\Psi_o(\mathbf{u}) \ldots = \Psi_\gamma(\mathbf{u}) = 0$, in virtue of Theorem 14 and Lemma 4 to the same, would have a non-zero solution in $k$, $\mathbf{u}^* = (u_o^*,\ldots,u_\mu^*)$. Since

$$h(\lambda) \prod_{\nu=1}^{n} (u_o^* + \rho_1^{(\nu)} u_1^* + \ldots \rho_\mu^{(\nu)} u_\mu^*) = 0$$

some factor $u_o^* + \rho_1^{(\nu)} u_1^* + \ldots + \rho_\mu^{(\nu)} u_\mu^* = 0$.
The same is true without the superscript $(\nu)$ and contradicts

the choice of $\rho_j$'s as a basis for a linear space over $k$.

Example:

Let $K = k(\lambda^{\frac{1}{\mu+1}})$,

$$\rho_i = \lambda^{\frac{i}{\mu+1}} \quad (1 \leq i \leq \mu),$$

$$F(x,\lambda) = N_{K/k(\lambda)}(\phi_o(\mathbf{x}) + \lambda^{\frac{1}{\mu+1}}\phi_1(\mathbf{x})+\ldots+\lambda^{\frac{\mu}{\mu+1}}\phi_\mu(\mathbf{x})).$$

$F$ is of degree $\mu$ in $\lambda$ which shows the theorem to be best possible.

Theorem 18. (Bertini 1882 for $k = \mathbb{C}$, Krull 1937 in general). *Let $k$ be an algebraically closed field and let $F(\boldsymbol{x},\boldsymbol{\lambda})$ be irreducible in $k(\lambda)$ and satisfy $\deg_{\boldsymbol{\lambda}} F = 1$. Then (1): $F(\boldsymbol{x},\lambda^*)$ is reducible in $k$ for every choice of $\boldsymbol{\lambda}^*$ with components in $k$ such that $\deg_{\boldsymbol{x}} F(\boldsymbol{x},\boldsymbol{\lambda}^*) = \deg_{\boldsymbol{x}} F(\boldsymbol{x},\boldsymbol{\lambda})$ if and only if either*

$$F(\boldsymbol{x},\lambda) = a_o(\lambda)\phi(\boldsymbol{x})^n + a_1(\lambda)\phi(\boldsymbol{x})^{n-1}\psi(\boldsymbol{x})+\ldots+a_n(\lambda)\psi(\boldsymbol{x})^n,$$

$\psi, \phi \in k[\boldsymbol{x}]$ *and* $\deg_{\boldsymbol{x}} F > \max(\deg \phi, \deg\psi)$

*or* $F(\boldsymbol{x},\boldsymbol{\lambda}) \in k(\boldsymbol{x}^p,\boldsymbol{\lambda})$, *where* $p = \operatorname{char} k$, $\boldsymbol{x}^p = \langle x_1^p, x_2^p, \ldots\rangle$.

Lemma 1. If under the assumptions of Theorem 18 $F(\mathbf{x},\boldsymbol{\lambda})$ has the property (1), then either it is reducible over $k(\boldsymbol{\lambda})^{sep}$ the separable closure of $k(\boldsymbol{\lambda})$, or $F(\mathbf{x},\boldsymbol{\lambda}) \in k[\mathbf{x}^p,\boldsymbol{\lambda}]$.

Proof. By Corollary 2 to Theorem 15 the hypothesis implies $F(\mathbf{x},\lambda)$ is reducible over $\widehat{k(\boldsymbol{\lambda})}$. Now $\widehat{k(\boldsymbol{\lambda})}$ is obtained from $k(\boldsymbol{\lambda})^{sep}$ by adjoining all elements $\alpha^{p^{-\nu}}$, $\alpha \in k(\boldsymbol{\lambda})^{sep}$

(Theorem D). Suppose $F(\mathbf{x},\lambda)$ is irreducible over $k(\lambda)^{sep}$ and that $I(\mathbf{x}) \mid F(\mathbf{x},\lambda)$, where $I(\mathbf{x})$ is irreducible over $k(\lambda)$. For a suitable $n$ $I(\mathbf{x})^{p^n}$ lies in $k(\lambda)^{sep}[\mathbf{x}]$ and since

$$I(\mathbf{x})^{p^n} \mid F(\mathbf{x},\lambda)^{p^n}$$

we have

$$I(\mathbf{x})^{p^n} = c_1 F(\mathbf{x},\lambda)^{\ell}, \quad c_1 \in k(\lambda)^{sep}.$$

On the other hand the irreducibility of $I(\mathbf{x})$ over $\widehat{k(\boldsymbol{\lambda})}$ implies

$$F(\mathbf{x},\lambda) = c_2 I(\mathbf{x})^m, \quad c_2 \in \widehat{k(\boldsymbol{\lambda})},$$

thus $\ell m = p^n$ and $m = p^r$. Furthermore $m > 1$ since $F(\mathbf{x},\lambda)$ in reducible over $\widehat{k(\lambda)}$ so $r > 0$ and $F(\mathbf{x},\lambda) = c_2 I(\mathbf{x})^{p^r} \in k(\mathbf{x}^p,\lambda)$, which proves the lemma.

Lemma 2. If under the assumption of Theorem 18 $F(\mathbf{x},\lambda)$ is reducible in $k(\lambda)^{sep}$ and $F \notin k[\mathbf{x}^p,\boldsymbol{\lambda}]$ then all monic factors of $F(\mathbf{x},\boldsymbol{\lambda})$ irreducible over $k(\lambda)^{sep}$ are conjugate over $k(\lambda)$ and each factor has at most two coefficients linearly independent over $k$.

Proof of Lemma 2. We proceed by induction on the number $\ell$ of components of $\boldsymbol{\lambda}$. If $\ell = 1$ we use Theorem 17. Assume now that the lemma holds for $\ell \leq \ell_o$. Let $\boldsymbol{\lambda}' = \langle\boldsymbol{\lambda},\mu\rangle$ have $\ell_o + 1$ components and

$$F \in k[\mathbf{x},\boldsymbol{\lambda},\mu] \setminus (k[\mathbf{x},\boldsymbol{\lambda}] \cup k[\mathbf{x},\mu]). \tag{2}$$

Since $\deg_{\boldsymbol{\lambda}'} F = 1$ the polynomial $F$ is irreducible over $\widehat{k(\lambda)}(\mu)$ and over $\widehat{k(\mu)}(\boldsymbol{\lambda})$.

Let $H(\mathbf{x})$ be a monic factor of $F$ defined and irreducible over $k(\boldsymbol{\lambda}')^{\mathrm{sep}}$ and $I(\mathbf{x})$ a monic factor of $H$ defined and irreducible over $\widehat{k(\boldsymbol{\lambda})}(\mu)^{\mathrm{sep}}$. Arguing as in the proof of Lemma 1 we get

$$H(\mathbf{x}) = I(\mathbf{x})^m, \quad \text{where} \quad m = \begin{cases} 1 & \text{if} \quad \mathrm{char}\, k = 0, \\ p^r & \text{if} \quad \mathrm{char}\, k = p, \end{cases}$$

thus $H \in k(\boldsymbol{\lambda}')^{\mathrm{sep}}[\mathbf{x}^m]$ and since $F$ is up to a factor from $k(\boldsymbol{\lambda},\mu)$ the product of all conjugates of $H$ with respect to $k(\boldsymbol{\lambda}')$

$$F \in k[x^m, \boldsymbol{\lambda}'].$$

By the assumption of the lemma $F \notin k[x^p, \boldsymbol{\lambda}']$, hence $m = 1$ and $H(\mathbf{x}) = I(\mathbf{x})$ is irreducible over $\widehat{k(\lambda)}(\mu)^{\mathrm{sep}}$. Similarly $H(\mathbf{x})$ is irreducible over $\widehat{k(\mu)}(\boldsymbol{\lambda})^{\mathrm{sep}}$. By Theorem 17 applied with $\widehat{k(\lambda)}$ in place of $k$ and $\mu$ in place of $\lambda$ we have

$$H(\mathbf{x}) = \phi(\mathbf{x}) + \rho\Psi(\mathbf{x}), \quad \text{where} \quad \phi, \Psi \in \widehat{k(\lambda)}[\mathbf{x}],$$

$$\rho \in \widehat{k(\boldsymbol{\lambda})}(\mu)^{\mathrm{sep}} \setminus \widehat{k(\boldsymbol{\lambda})}.$$

For $\rho$ we take the coefficient of any monomial occuring in $H(\mathbf{x})$ that does not belong to $\widehat{k(\boldsymbol{\lambda})}[\mathbf{x}]$. Such monomials exist since otherwise $H(\mathbf{x})$ would divide both the coefficient of $\mu$ in $F(\mathbf{x},\boldsymbol{\lambda},\mu)$ and $F(\mathbf{x},\boldsymbol{\lambda},0)$ and these two polynomials would have a non-trivial common factor over $k(\boldsymbol{\lambda})$, which in view of (2) contradicts irreducibility of $F$ over $k(\boldsymbol{\lambda},\mu)$. On the other hand $H \in k(\boldsymbol{\lambda},\mu)^{sep}[\mathbf{x}]$, hence

$$(3) \qquad \phi,\psi \in k(\boldsymbol{\lambda})^{sep}[\mathbf{x}] \ , \ \rho \in k(\boldsymbol{\lambda},\mu)^{sep} \setminus k(\boldsymbol{\lambda})^{sep} .$$

Moreover by the choice of $\rho$

(4) a certain monic monomial $M$ occurs in $\psi(\mathbf{x})$, but not in $\phi(\mathbf{x})$.

Take conjugates $H_\nu = \phi_\nu(\mathbf{x}) + \rho_\nu\psi_\nu(\mathbf{x})$ of $H$ over $k(\boldsymbol{\lambda},\mu)$. They also divide $F(\mathbf{x},\boldsymbol{\lambda},\mu)$ and are irreducible over $\widehat{k(\boldsymbol{\lambda})}(\mu)$. By Theorem 17 for each $\nu$ there is a $\rho'_\nu \in \widehat{k(\boldsymbol{\lambda})}(\mu)^{sep}$ such that

$$(5) \qquad \phi_\nu(\mathbf{x}) + \rho_\nu\psi_\nu(\mathbf{x}) = \phi(\mathbf{x}) + \rho'_\nu\psi(\mathbf{x}).$$

The monomial $M$ mentioned in (4) occurs in $\psi_\nu(\mathbf{x})$, with coefficient 1, but in $\phi_\nu(\mathbf{x})$ with coefficient 0. Comparing the coefficient of $M$ on both sides of (5) we get $\rho_\nu = \rho'_\nu$ and since by (3) $\phi_\nu,\psi_\nu \in k(\boldsymbol{\lambda})^{sep}[\mathbf{x}]$, $\rho_\nu \notin k(\boldsymbol{\lambda})^{sep}$ we get

$$\phi_\nu(\mathbf{x}) = \phi(\mathbf{x}), \quad \psi_\nu(\mathbf{x}) = \psi(\mathbf{x})$$

Since this holds for all conjugates $H_\nu$ of $H$ over $k(\boldsymbol{\lambda},\mu)$ we have

$$\phi, \psi \in k(\boldsymbol{\lambda})[\mathbf{x}].$$

Moreover $\rho \notin \widehat{k(\mu)}$ since otherwise $F$ would have a factor over $\widehat{k(\mu)}(\boldsymbol{\lambda})$. Now we apply the inductive assumption with $\widehat{k(\mu)}$ in place of $k$ and get

$$H(\mathbf{x}) = \phi'(\mathbf{x}) + \rho\psi'(\mathbf{x}),$$

where $\phi', \psi' \in k(\mu)^{sep}[\mathbf{x}]$. The same argument as before leads to the conclusion that $\phi', \psi' \in k(\mu)[\mathbf{x}]$. We have

$$\phi(\mathbf{x}) + \rho\psi(\mathbf{x}) = \phi'(\mathbf{x}) + \rho\psi'(\mathbf{x})$$

and unless $\psi = \psi'$ we get

$$\rho = \frac{\phi'-\phi}{\psi-\psi'} \in k(\mathbf{x}, \boldsymbol{\lambda}, \mu).$$

But $\rho$ does not depend on $\mathbf{x}$, thus $\rho \in k(\boldsymbol{\lambda}, \mu)$. Since $F(\mathbf{x}, \boldsymbol{\lambda}, \mu)$ is irreducible over $k(\boldsymbol{\lambda}, \mu)$ this is impossible and we get $\psi = \psi'$, $\phi = \phi'$. Hence

$$\phi, \psi \in k(\boldsymbol{\lambda})[\mathbf{x}] \cap k(\mu)[\mathbf{x}] = k[\mathbf{x}]$$

This proves the lemma.

<u>Proof of Theorem 18</u>. The condition given in the theorem as equivalent to (1) is necessary. Indeed by Lemma 1 and 2 if (1) holds then either $F(\mathbf{x}, \boldsymbol{\lambda}) \in k(\mathbf{x}^p, \boldsymbol{\lambda})$, $p = \text{char } k$ or

$$F(\mathbf{x},\lambda) = h(\boldsymbol{\lambda}) \prod_{\nu=1}^{n} (\phi(x) + \rho^{(\nu)}\psi(X)),$$

where $n > 1$, $\phi(\mathbf{x})$, $\psi(x) \in k[x]$ and $\rho^{(\nu)}$ are conjugate over $k(\boldsymbol{\lambda})$. In the latter case we have

$$f(\mathbf{x},\lambda) = a_0(\boldsymbol{\lambda})\phi(\mathbf{x})^n + a_1(\boldsymbol{\lambda})\phi(x)^{n-1}\psi(x)+\ldots+a_n(\boldsymbol{\lambda})\psi(x)^n$$

and moreover

$$\deg_{\mathbf{x}} F(\mathbf{x},\lambda) = \sum_{\nu=1}^{n} \deg_{\mathbf{x}}(\phi(\mathbf{x}) + \rho^{(\nu)}\psi(\mathbf{x}))$$

$$= n \max\{\deg \phi, \deg \psi\} > \max(\deg \phi, \deg \psi\}.$$

On the other hand if the condition given in the theorem and demonstrated above is fullfilled then either for every choice of $\lambda^*$, $F(x,\lambda^*)$ is a $p^{th}$ power or for every choice of $\lambda^*$ that does not diminish $\deg_{\mathbf{x}} F(\mathbf{x},\lambda)$, $F(\mathbf{x},\lambda^*)$ is a product of factors of smaller degree.

Remark. The above proof works in the case $|F|_{\boldsymbol{\lambda}} = \max_{1\le i\le \ell} |F|_{\lambda_i} = 1$ and Theorem 18 remains true in this case. Still further generalization has been claimed by Riehle 1919, but Krull 1937 objects to the validity of his proof. For a discussion of Bertini's theorem from the point of view of algebraic geometry see van der **Waerden** 1937, Zariski 1941 and Segre 1946.

## Section 12. Mertens's theorems on reducibility

Theorem 19 (Mertens 1911) *For arbitrary positive integers* $n$ *and* $m$, $n>m$ *there exists an integral polynomial* $F(z,u_1,..,u_m,c_1,..,c_n)$ *monic with respect to* $z$ *with the following properties*

*(i) If* $disc_z F(z,u_1^*,..,u_m^*,c_1^*,...,c_n^*) \neq 0$ *for some* $u_1^*,...,u_m^*,c_1^*,..,c_n^* \in k$

*then the polynomial* $f^*(x) = x^n + \sum_{i=1}^{n} c_i^* x^{n-i}$ *has a factor of degree* $m$ *in* $k$ *if and only if* $F(z,u_1^*,..,u_m^*,c_1^*,..,c_n^*)$ *has a zero in* $k$.

*(ii) If* $disc\ f^* \neq 0$ *then in every infinite subset of* $k$ *there exist elements* $u_1^*,..,u_m^*$ *such that* $disc_z\ F(z,u_1^*,..,u_m^*,c_1^*,..,c_n^*) \neq 0$.

Proof. Let $x_1,..,x_n$ be indeterminates, let $c_j$ be the jth symmetric function in the $x_i (1 \leq i \leq n)$. If $\omega$ is a subset of $\{1,..,n\}$ of cardinality $m$, let $\tau_{j\omega}$ be the jth symmetric function of the $x_i$ $(i \in \omega)$. Let $\Omega$ be the family of all subsets of $\{1,..,n\}$ of cardinality $m$. Define

$$F(z,u_1,..,u_m,c_1,..,c_n) = \prod_{\omega \in \Omega} (z-u_1\tau_{1\omega} -..-u_m\tau_{m\omega}).$$

Since every permutation on $\{x_1,..,x_n\}$ permutes the elements of $\Omega$ the expression on the right is in $\mathbb{Z}[z,\mathbf{u},\mathbf{c}]$, where

$\mathbf{u} = \langle u_1,..,u_m \rangle$, $\mathbf{c} = \langle c_1,..,c_n \rangle$. Let

$$f(x) = x^n + c_1 x^{n-1} + ... + c_n = \prod_{j=1}^{n} (x + x_j),$$

$$f^*(x) = \prod_{j=1}^{n} (x+x_j^*) \ , \quad \text{where} \quad x_j^* \in k$$

and suppose $g^* \in k[x]$ , $\deg g^* = m$ and $g^* | f^*$ . Then for some $\omega \in \Omega$ and $c \in k$

$$g^*(x) = c \prod_{j \in \omega} (x+x_j^*) = c(x^m + \tau_{1\omega}^* x^{m-1} + .. + \tau_{m\omega}^* )$$

and for every choice of $u_1^*, .., u_m^* \in k$ the element $u_1^* \tau_{1\omega}^* + u_m^* \tau_{m\omega}^*$ of $k$ is a zero of $F(z, u_1^*, .., u_m, c_1^*, .. c_n^*)$.

On the other hand, let

$$F_o = \frac{\partial F}{\partial z} \ , \quad F_i = - \frac{\partial F}{\partial u_i} \qquad (1 \leq i \leq n)$$

For every $\omega \in \Omega$ we have

$$F( u_1 \tau_{1\omega} + ... + u_m \tau_{m\omega}, \mathbf{u}, \mathbf{c}) = 0$$

identically in $\mathbf{u}$. Hence if we differentiate this identity with respect to $u_i$ we obtain

$$\tau_{i\omega} F_o(u_1 \tau_{1\omega} + .. + u_m \tau_{m\omega}, \mathbf{u}, \mathbf{c}) - F_i(u_1 \tau_{1\omega} + .. + u_m \tau_{m\omega}, \mathbf{u}, \mathbf{c}) = 0$$

so that

$$\tau_{i\omega} = \frac{F_i(u_1 \tau_{1\omega} + .. + u_m \tau_{m\omega} \ \mathbf{u}, \ \mathbf{c})}{F_o(u_1 \tau_{1\omega} + .. + u_m \tau_{m\omega}, \mathbf{u}, \ \mathbf{c})} \ .$$

Define

$$P(x) = F_o(z, \mathbf{u}, \mathbf{c}) \ x^m + F_1(z, \mathbf{u}, \mathbf{c}) \ x^{m-1} + .. + F_m(z, \mathbf{u}, \mathbf{c})$$

Then

$$F_o^{n-m+1} f(x) = PQ + R \ , \ \text{where } P, Q, R \in k[x, z, \mathbf{u}, \mathbf{c}], \quad \deg_x R < m \tag{1}$$

and

$$F_o^{n-m+1}(u_1\tau_{1\omega}+..+u_m\tau_{m\omega},\mathbf{u},\mathbf{c})f(x)=F_o(u_1\tau_{1\omega}+..+u_m\tau_{m\omega},\mathbf{u},\mathbf{c}).$$

$$[x^m+\tau_{1\omega}x^{m-1}+..+\tau_{m\omega}]Q+R.$$

But $x^m+\tau_{1\omega}x^{m-1}+...+\tau_{m\omega}$ divides $f(x)$ and so divides R, whence

$$R(x,u_1\tau_{1\omega}+...+u_m\tau_{m\omega}\ ,\ \mathbf{u}\ ,\ \mathbf{c})\ =\ 0$$

identically in $u_1,..,u_m,x_1,..,x_n$. Consequently

$$z-u_1\tau_{1\omega}-...-u_m\tau_{m\omega}\ \Big|\ R(x,z,\mathbf{u},\mathbf{c})$$

in the ring $Z[x,z,\mathbf{u},x_1,..,x_n]$ and since this is true for all sets $\omega\in\Omega$ we have

$$F(z,\mathbf{u},\mathbf{c})\ |\ R(x,z,\mathbf{u},\mathbf{c}).$$

Thus $R(x,z,\mathbf{u},c) = F(z,\mathbf{u},\mathbf{c})\ G(x,z,\mathbf{u},\mathbf{c})$,

where G as symmetric in the $x_i$ is in $\mathbf{Z}[x,z,\mathbf{u},\mathbf{c}]$.
Now let $\mathbf{u}^*\in k^m, \mathbf{c}^*\in k^n, z^*\in k$ satisfy

(2) $\quad \mathrm{disc}_z\ F(z,\mathbf{u}^*,\mathbf{c}^*) \neq 0$ and $F(z^*,\mathbf{u}^*,\mathbf{c}^*) = 0$

Then $R(x,z^*,\mathbf{u}^*,\mathbf{c}^*) = 0$, whence by (1) $P(x,z^*,\mathbf{u}^*,\mathbf{c}^*)$ divides $F_o^{n-m+1}(z^*,\mathbf{u}^*,\mathbf{c}^*)f^*(x)$, i.e. $f^*$ has a factor of degree m with coefficients in k provided $F_o(z^*,\mathbf{u}^*,\mathbf{c}^*) \neq 0$.
Since however $F_o(z^*,\mathbf{u}^*,\mathbf{c}^*) = \frac{\partial}{\partial z}F(z,\mathbf{u}^*,\mathbf{c}^*)$ the latter condition is implied by (2). This proves (i). In order to

prove (ii) observe that if for a given $f^*$ hence for given $x_1^*, \ldots x_n^*$

(3) $\operatorname{disc}_z F(z,\mathbf{u}^*,\mathbf{c}^*) =$

$$\prod_{\substack{\omega,\omega'\in\Omega \\ \omega\neq\omega'}} (u_1^* \tau_{1\omega}^* + \ldots + u_m^* \tau_{m\omega}^* - u_1^* \tau_{1\omega'}^* - \ldots - u_m \tau_{m\,\omega'}^*)$$

is zero for every choice of $u_1^*, \ldots, u_m^*$ in an infinite subset of $k$ then the product

$$\prod_{\substack{\omega,\omega'\in\Omega \\ \omega\neq\omega'}} (u_1 \tau_{1\omega}^* + \ldots + u_m \tau_{m\omega}^* - u_1 \tau_{1\omega'}^* - \ldots - u_m \tau_{m\omega'}^*)$$

has a zero factor. Hence for some $\omega \neq \omega'$ we have $\tau_{i\omega} = \tau_{i\omega'}$ $(1 \leq i \leq m)$ so that the sequence $\{x_j^*\}_{j\in\omega}$ is a permutation of $\{x_j^*\}_{j\in\omega'}$. Since $\omega \neq \omega'$ we must have two $x_j^*$'s equal and hence $\operatorname{disc} f = 0$.

The proof of Theorem 19 is complete. The theorem will be used in § 22.

## Section 13. Capelli's theorem

We recall that $F(x_1,\dots,x_s) \overset{can}{\underset{k}{=}} \text{const} \prod_{\rho=1}^{r} F_\rho(x_1,\dots,x_s)^{e_\rho}$ means that the $F_\rho$ are irreducible over k and prime to each other.

Theorem 20. (Capelli 1897 for char k=0). *Let G be irreducible over k, $G(\beta)=0$. If $H(x)-\beta \overset{can}{\underset{k(\beta)}{=}} \text{const} \prod_{\rho=1}^{r} \Phi_\rho(x)^{e_\rho}$, then $G(H(x)) \overset{can}{\underset{k}{=}} \text{const} \prod_{\rho=1}^{r} N_{k(\beta)/k}\Phi_\rho(x)^{e_\rho}$.*

Proof. (i) G separable. Here $\beta^{(\nu)}$, the conjugates of $\beta$ over k are distinct. We want to show that the polynomials $N_{k(\beta)/k}\Phi_\rho(x)$ are irreducible over k.

If

$$\Phi_\rho(x) \mid I_\rho(x), \quad I_\rho \text{ irreducible in } k,$$

then

$$\Phi_\rho^{(\nu)}(x) \mid I_\rho(x).$$

We have

$$(1) \qquad \Phi_\rho(x) \mid H(x)-\beta \quad \text{so} \quad \Phi_\rho^{(\nu)}(x) \mid H(x)-\beta^{(\nu)}.$$

$\beta \neq \beta^{(\nu)}$ implies that $\Phi_\rho, \Phi_\rho^{(\nu)}$ are relatively prime. Hence

$$N_{k(\beta)/k}\Phi_\rho(x) \mid I_\rho(x)$$

and $N_{k(\beta)/k}\Phi_\rho(x) = \text{const } I_\rho(x)$ is irreducible.

We need yet that norms are coprime.

Now $(\Phi_1,\Phi_2)=1$ implies in virtue of (1) $(\Phi_1,\Phi_2^{(j)})=1$,

so $(\Phi_1, N_{k(\beta)/k}\Phi_2)=1$, and $(N_{k(\beta)/k}\Phi_1, N_{k(\beta)/k}\Phi_2)=1$.

(ii). G purely inseparable. Here

$$G(x) = x^{p^{\nu}} - a\ ,\quad a \in k\ ,\quad a^{p^{-1}} \notin k\ ,\quad \beta = a^{p^{-\nu}} \quad \text{and}$$

$$N_{k(\beta)/k}\, \Phi_\rho(x) = \Phi_\rho^{\,p^{\nu}}.$$

If the above polynomial is not irreducible in $k$ then $\Phi_\rho^{p^{\nu-1}}$ lies in $k[x]$, and since

$$\Phi_\rho^{\,p^{\nu-1}} \,|\, H(x)^{p^{\nu-1}} - a^{p^{-1}}\ ,$$

we have $a^{p^{-1}} \in k$ by the Euclidean algorithm, a contradiction.

(iii). General case. We can write $G = G_1(x^{p^{\nu}})$ with $G_1$ separable. We start with $G_1(H(x)^{p^{\nu}})$. If $G(\beta) = 0$ then $G_1(\beta^{p^{\nu}}) = 0$.

(ii) gives $H(x)^{p^{\nu}} - \beta^{p^{\nu}} \underset{k(\beta^{p^{\nu}})}{\overset{\text{can}}{=}} \text{const} \prod_{\rho=1}^{r} N_{k(\beta)/k(\beta^{p^{\nu}})} \Phi_\rho^{e_\rho}$

so
$$G_1(H(x)^{p^{\nu}}) \underset{k}{\overset{\text{can}}{=}} \text{const} \prod_{\rho=1}^{r} N_{k(\beta^{p^{\nu}})/k} \left(N_{k(\beta)/k(\beta^{p^{\nu}})} \Phi_\rho^{e_\rho}\right)$$

$$= \text{const} \prod_{\rho=1}^{r} N_{k(\beta)/k} \Phi_\rho^{e_\rho}$$

from (1).

Corollary 1. If $F = G_1(H_1(x)) = G_2(H_2(x))$ and the degrees

$|G_1| = |H_2|, |H_1| = |G_2|$ are coprime, then $F$ is irreducible in $k$ if and only if $G_1$ and $G_2$ are irreducible in $k$.

Proof. Each irreducible factor has degree divisible by that of $G_1$ and $G_2$, hence by $|F|$.

Corollary 2. If $n = \prod_{i=1}^{j} p_i^{\alpha_i}$ then $x^n - a$ is irreducible in $k$ if and only if $x^{p_i^{\alpha_i}} - a$ is irreducible in $k$ for all $i \le j$.

<u>Proof.</u> $$x^n - a = (x^{p_j^{\alpha_j}})^{p_1^{\alpha_1} \cdots p_{j-1}^{\alpha_{j-1}}} - a.$$

We use Corollary 1 and induction on $j$.

<u>Theorem 21</u>. (Capelli 1898 for char $k=0$, Rédei 1959 in general) *$x^n-a$ is reducible in $k$ if and only if either for some prime divisor $p$ of $n$ $a=b^p$, $b \in k$, or $4|n$ and $a=-4b^4$, $b \in k$.*

<u>Lemma 1</u>. (Abel) If $p$ is a prime $x^p-a$ is reducible in $k$ if and only if $a=b^p$, $b\in k$.

<u>Proof</u>. If $g(x)|x^p-a$ with $p>|g|>0$ and $g$ is monic then the constant term of $g$ is a product of $|g|$ terms of the form $\zeta_p^i a^{1/p}$, whence $\zeta_p^j a^{|g|/p} = c\in k$ thus $c^p= a^{|g|}$ and $a = b^p$ since $(p,|g|) = 1$. The converse implication is obvious.

<u>Lemma 2.</u> Theorem 21 holds for $n=p^\nu$ $(p>2)$.

<u>Proof</u>. We proceed by induction on $\nu$. The case $\nu = 1$ is Lemma 1. Suppose $\nu \geq 2$ and the result holds for $\nu-1$. Now if $a \neq b^p$

$$x^{p^\nu} - a = (x^{p^{\nu-1}})^p - a \quad \text{is reducible in } k$$

only if $x^{p^{\nu-1}} - a^{1/p}$ is reducible in $k(a^{1/p})$ by Theorem 20. (Here $G = x^p-a$ and is irreducible by Lemma 1).

The inductive assumption gives

$$a^{1/p} = (\Omega(a^{1/p}))^p, \ \Omega \in k[x].$$

On taking norms

$$N_{k(a^{1/p})/k}\, a^{1/p} = a = (N_{k(a^{1/p})/k}\, \Omega(a^{1/p}))^p$$

so a is a p-th power in k, a contradiction. Thus $a \neq b^p$ implies $x^{p^\nu} - a$ irreducible. The converse implication is obvious.

Lemma 3. Theorem 21 holds for $n = 2^\nu$.

Proof. $x^4 + 4b^4 = (x^2-2bx + 2b^2)(x^2 + 2bx + 2b^2)$ gives sufficiency of the condition. To prove the necessity we use induction and assume $a \neq b^2$ in k. By Theorem 20 and Lemma 1

$$x^{2^\nu} - a = (x^{2^{\nu-1}})^2 - a \quad \text{is reducible in k only if}$$

$x^{2^{\nu-1}} - \sqrt{a}$ is reducible in $k(\sqrt{a})$.

By the inductive assumption either $\sqrt{a} = (c + d\sqrt{a})^2$ or

$$\sqrt{a} = -4(c+d\sqrt{a})^4 \quad \text{and} \quad \nu \geq 3 .$$

The first case implies $\sqrt{a} = c^2 + 2cd\sqrt{a} + d^2a$, and we therefore have (since $a \notin k^2$)

$$c^2 + ad^2 = 0 , \quad \text{and} \quad 2cd = 1 .$$

Hence $a = -\dfrac{c^2}{d^2} = -4c^4$ .

The second case implies

$$\sqrt{a} = -4(c^4 + 6c^2d^2a + d^4a^2) - 16\sqrt{a}\,(c^3d + cd^3a) ,$$

so

$$c^4 + 6c^2d^2a + d^4a^2 = 0 , \quad \text{whence } 4c^2d^2a = -(d^2a + c^2)^2$$
$$\text{and} \quad 1 = -16cd(d^2a + c^2) .$$

Hence

$$a = -4\left(\frac{1}{8cd}\right)^4 ,$$ which completes the induction.

Proof of Theorem 21. The theorem follows from Lemma 2 and 3 and the Corollary 2 to Theorem 20.

Corollary 1. Let $p = \text{char } k$ and $G(x)$ be monic and irreducible over k. $G(x^{p^\nu})$ is reducible over $k$ if and only if all coefficients of $G$ are $p^{th}$ powers in k.

Proof. Let $G(\beta) = 0$ . By Theorems 20 and 21 if $G(x^{p^\nu})$ is reducible in $k$ then $\beta = \gamma^p$ , where $\gamma \in k(\beta)$.

Hence

$$x^{p^\nu} - \beta = (x^{p^{\nu-1}} - \gamma)^p$$

and again by Theorem 20 $G(x^{p^\nu}) = \text{const } H(x)^p$, $H \in k[x]$. Since $G$ is monic it follows that all coefficients of $G$ are p-th powers in $k$. Conversely if this is the case $G(x^{p^\nu})$ is reducible in $k$ .

Corollary 2 to Theorem 21. If $F(x_1,\dots,x_r,\dots,x_s)$ is irreducible over a field $k$ of characteristic $p$ and

$F(x_1^{p^{\nu_1}},\dots,x_r^{p^{\nu_r}},x_{r+1},\dots,x_s)$ is reducible over $k$ then $F(x_1,\dots,x_r,\dots,x_s) = G(x_1,\dots,x_r,x_{r+1}^p,\dots,x_s^p)$.

Proof. It is enough to prove the corollary for $r=\nu_1=1$, the general case follows then by induction on $\nu_1+\dots+\nu_r$. Let $F(x_1,x_2,\dots,x_s) = \sum_{i=0}^{j} a_i(x_2,\dots,x_s)x_1^{j-i}$. Since $F$ is irreducible in $k$ $(a_o,\dots,a_j)=1$. Hence reducibility of $F(x_1^p,x_2,\dots,x_s)$ over $k$ implies its reducibility over $k(x_2,\dots,x_s)$. By Corollary 1 all quotients $a_i(x_2,\dots,x_s)a_o^{-1}(x_2,\dots,x_s)$ $(i\le j)$ are $p^{th}$ powers in $k(x_2,\dots,x_s)$. Since $(a_o,\dots,a_j)=1$ it follows that $a_i(x_2,\dots,x_s)= c_ib_i(x_2,\dots,x_s)^p$, $c_i$ in $k$, and we can take $G(x_1,x_2,\dots,x_s) = \sum_{i=0}^{j} c_iB_i(x_2,\dots,x_s)x_1^{j-i}$, where $B_i$ is obtained from $b_i$ by the endomorphism $k \longrightarrow k^p$.

Corollary 3 to Theorem 21. (Ehrenfeucht, 1955). If $(|F|,|G|)=1$ then $F(x) - G(y)$ is irreducible.

Proof. If reducible, $F(x) - G(y) = H_1(x,y)\,H_2(x,y)$. Let the weight of $x$ be $|G|$, that of $y$ be $|F|$, and let the leading coefficients of $F$ and $G$ be $f$ and $g$ respectively. The highest isobaric part of $F(x)-G(y)$ is the product of the highest isobaric parts of $H_1,H_2$; hence

$$f\,x^{|F|} - g\,y^{|G|} \text{ is reducible.}$$

By Capelli's theorem applied with $\hat{k}(y)$ in place of $k$ $f^{-1}\,g\,y^{|G|}$ is a power with prime exponent dividing $|F|$, a contradiction.

Theorem 22. (Schinzel 1973b). *Let* $F(x_1,\ldots,x_l) = \sum_{j=0}^{h} a_j \prod_{i=1}^{l} x_i^{\nu_{ij}}$, *where* $a_j \neq 0$ $(0 \leq j \leq h)$ *and the vectors* $(\nu_{1j},\ldots,\nu_{lj})$ *are all distinct for* $j = 0,\ldots,h$. *If rank of the matrix*

$$M = \begin{pmatrix} 1 & \cdots\cdots\cdots & 1 \\ \nu_{1o} & \cdots\cdots\cdots & \nu_{1h} \\ \vdots & & \vdots \\ \nu_{lo} & \cdots\cdots\cdots & \nu_{lh} \end{pmatrix}$$

*over the prime field of* $k$ *is greater than* $\frac{h+3}{2}$ *then* $F \prod_{i=1}^{l} x_i^{-\min_j \nu_{ij}}$ *is irreducible over* $k$.

Proof. First we recall some notation. If a rational function $\phi$ satisfies

$$\phi(x_1,\ldots,x_\ell) = \prod_{i=1}^{\ell} x_i^{\alpha_i}\, G(x_1,\ldots,x_\ell), G \in k[x_1,\ldots x_2],$$

where $(G(x_1,\ldots,x_\ell), x_1\ldots x_\ell) = 1$, then we write $J\phi(x_1,\ldots,x_\ell) = G(x_1,\ldots,x_\ell)$, so in the statement of the theorem we can replace $F \prod_{i=1}^{\ell} x_i^{-\min \nu_{ij}}$ by $JF$.

We now proceed by induction on $\ell$. The case $\ell = 1$ is vacuous. Assume the theorem true for less then $\ell$ variables. Without loss of generality assume the minor of the upper left-hand corner of the matrix $M$ is nonvanishing and is of maximal possible size with that property say of size $r+1$. It follows that

$$D = \det_{1 \leq i,j \leq r} (\nu_{ij} - \nu_{io}) \not\equiv 0 \bmod \text{char } k.$$

Assume moreover interchanging two variables if necessary that $D > 0$. Let $(\alpha_{ij})$ denote the matrix $(\nu_{ij} - \nu_{io})^{-1}_{1\le i,j\le r}$.

Make the change of variables:

$$(1) \qquad x_q = \begin{cases} \prod_{s=1}^{r} \left(y_s \prod_{t=r+1}^{\ell} y_t^{-\nu_{ts}+\nu_{t0}}\right)^{D\alpha_{s,q}} & \text{if } q \le r \\ y_q^D & \text{if } q > r. \end{cases}$$

We get

$$F(x_1,\dots,x_\ell) \prod_{i=1}^{\ell} x_i^{-\nu_{io}} = \sum_{j=0}^{h} a_j \prod_{i=1}^{\ell} y_i^{\lambda_{ij}}, \text{ where}$$

$$\lambda_{ij} = \begin{cases} D \sum_{q=1}^{r} \alpha_{iq} (\nu_{qj} - \nu_{qo}) & \text{if } i \le r, \\ D \sum_{q,s=1}^{r} \alpha_{sq} (\nu_{qj} - \nu_{qo})(-\nu_{is} + \nu_{io}) + D(\nu_{ij} - \nu_{io}) & \text{if } i > r. \end{cases}$$

By the definition of $(\alpha_{ij})$ and the fact that $(\nu_{ij} - \nu_{io})$ is of rank $r$ we have

$$\lambda_{ij} = \begin{cases} \delta_{ij} D & \text{if both } i \le r \text{ and } j \le r, \\ 0 & \text{if } i > r. \end{cases}$$

Furthermore, since the transformation (1) considered as a linear transformation of the logarithms of variables $x_i, y_i$

is non-singular, it transforms distinct monic monomials in the $x_i$ into distinct monic monomials in the $y_i$, hence the vectors $\langle \lambda_{1j},\dots,\lambda_{rj} \rangle$ are all distinct ( $0\le j\le h$).

Moreover

$$J\Big(\sum_{j=0}^{h} a_j \prod_{i=1}^{\ell} y_i^{\lambda_{ij}}\Big) = \Big(\sum_{j=0}^{h} a_j \prod_{i=1}^{r} y_i^{\lambda_{ij}}\Big) \prod_{i=1}^{r} y_i^{-\min_j \lambda_{ij}}$$

is reducible if JF is.

Now

$$\sum_{j=0}^{h} a_j \prod_{i=1}^{r} y_i^{\lambda_{ij}} = a_o + a_1 y_1^D + a_2 y_2^D + \dots + a_r y_r^D + \sum_{j=r+1}^{h} a_j \prod_{i=1}^{r} y_i^{\lambda_{ij}},$$

$$= (a_o + a_1 y_1^D + \dots) + y_2^D(a_2 + \dots) + \dots + y_r^D(a_r + \dots) + \dots$$

hence

(2) G has no factor in $k[y_1]$.

Indeed in the opposite case since there are above at least r monomials in $y_2,\dots,y_r$ the total number of terms in G would be at least $2r>h+1$.

The reducibility of G implies

(3) there exists a $\sigma>r$ such that $\sum_{i=2}^{r} \lambda_{i\sigma} \neq D$.

Indeed if this does not hold we make the change of variables

$$y_3 = y_2 z_3, \dots, y_r = y_2 z_r$$

and get

$$\sum_{j=0}^{h} a_j \prod_{i=1}^{r} y_i^{\lambda_{ij}} = a_o + a_1 y_1^D + y_2^D \Big(a_2 + a_3 z_3^D + \dots + a_r z_r^D$$

$$+ \sum_{j=r+1}^{h} a_j y_1^{\lambda_{1j}} \prod_{i=3}^{r} z_i^{\lambda_{ij}}\Big)$$

If $G$ is reducible then either the above function viewed as a binomial in $y_2$ is reducible over $k(y_1, z_3, \ldots, z_r)$ or the polynomials $a_o + a_1 y_1^D$ and $J\,(a_2 + a_3 z_3^D + \ldots + a_r z_r^D + \sum_{j=r+1}^{h} a_j y_1^{\lambda_{ij}} \prod_{i=3}^{r} z_i^{\lambda_{ij}})$ have a common factor.

Such a factor would have to belong to $k[y_1]$ and to **divide** $G$, **hence** the latter situation is impossible by (2). Applying Theorem 21 for the field $\hat{k}(y_1, z_3, \ldots, z_r)$ we infer that

$$\frac{a_o + a_1 y_1^D}{a_2 + a_3 z_3^D + \ldots + a_r z_r^D + \sum_{j=r+1}^{h} a_j y_1^{\lambda_{1j}} \prod_{i=3}^{r} z_i^{\lambda_{ij}}}$$

is a power in $\hat{k}(y_1, z_3, \ldots, z_r)$, hence $a_o + a_1 y_1^D$ is a power in $\hat{k}(y_1)$. Since $D \not\equiv 0 \mod \operatorname{char} k$, this is impossible and the obtained contradiction proves (3).

We distinguish two cases:

(i) there exist distinct indices $p > 0$ and $q > 0$ such that

$$\lambda_{ip} = \lambda_{iq} \quad \text{for all} \quad i > 1,$$

(ii) for every two distinct indices $p > 0$ and $q > 0$ there exists an $i > 1$ such that

$$\lambda_{ip} \neq \lambda_{iq}.$$

In case (i) divide the indices $j$ into classes in the following way: $j_1, j_2$ are in class $C_s$ $(0 \leq s \leq t)$ if $\lambda_{ij_1} = \lambda_{ij_2} = \mu_{is}$ for all $i > 1$. Clearly the vectors

$< \mu_{2s},\dots,\mu_{rs} >$ are all distinct and the polynomials $A_s(y_1) = \sum_{j\in C_s} a_j y_1^{\lambda_{ij}}$ are all different from $0 (0 \le s \le t)$.

Moreover

$$(4) \qquad H = \sum_{s=0}^{t} A_s(y_1) \prod_{i=2}^{r} y_i^{\mu_{is}} = \sum_{j=0}^{h} a_j \prod_{i=1}^{r} y_i^{\lambda_{ij}} =$$

$$= G(y_1,\dots,y_\ell) \prod_{i=1}^{r} y_i^{\min \lambda_{ij}} .$$

Consider $JH$ as a polynomial over $k(y_1)$. Since $(0,\dots,0)$, $(D,\dots,0),\dots,(0,\dots,D)$ occur among the vectors $<\mu_{2s},\dots,\mu_{rs}>$ $(0 \le s \le t)$ the rank of the matrix $\begin{pmatrix} 1 \dots 1 \\ \mu_{is} \end{pmatrix}$ is $r$.

On the other hand, by (i) $t \le h-2$. Hence $r > \frac{t+3}{2}$ and by the inductive assumpstion the polynomial $JH$ is irreducible over $k(y_1)$. By (4) the same applies to $G$, hence if $G$ is reducible over $k$ it has a factor in $k[y_1]$ contrary to (2).

In case (ii) make the substitution

$$y_1 = \left( - \frac{a_0}{a_1} \right)^{1/D} = \eta .$$

If

$$G(y_1,y_2,\dots,y_r) = G_1(y_1,y_2,\dots,y_r) G_2(y_1,y_2,\dots,y_r), G_i \notin k$$

we get

$$(5) \qquad G(\eta,y_2,\dots,y_r) = G_1(\eta,y_2,\dots,y_r) \, G_2(\eta,y_2,\dots,y_r).$$

By the condition (ii) there is no cancellation of terms of positive degree in $y_2,\dots,y_r$ on the left hand side of (5), which must therefore be reducible.

This contradicts the inductive assumption for the field $k(\eta)$ since the number of terms on the left is $h-2$ and by (3) the rank of the relevant matrix is $r$.

Remark 1. Theorem 22 holds without the assumption that vectors $\langle \nu_{1j},\ldots,\nu_{\ell j}\rangle$ are distinct. If not, we group similar terms to obtain $\sum_{j=i}^{m} b_j \prod_{i=1}^{\ell} x_i^{\mu_{ij}}$, where

$\langle \mu_{1j},\ldots,\mu_{\ell j}\rangle$ are distinct. Denoting the new rank by $r'$, we have $r-r' \le \frac{h-m}{2}$ hence if $r+1>\frac{m+3}{2}$ then $r'+1>\frac{m+3}{2}$ and the theorem applies.

2. Theorem 22 is best possible, as the following example shows.

$$F(x_1,\ldots,x_\ell)=(1+x_1)(1+\sum_{i=2}^{\ell} x_i)=1+\sum_{i=1}^{\ell} x_i + x_1\sum_{i=2}^{\ell} x_i .$$

The matrix in question looks like

$$\begin{pmatrix} 1\,1\ldots & \ldots 1 \\ 0 & 1\,1\ldots 1 \\ \vdots \quad I_\ell & \\ \vdots & I_{\ell-1} \\ 0 & \end{pmatrix}$$

and its rank $\ell+1 = \frac{h+3}{2}$ since $h = 2\ell-1$.

3. If $r=h=3$ and char $k=0$ there are three types of reducible polynomials $F(x_1,\ldots,x_\ell)$ prime to $x_1\ldots x_\ell$ namely

$$U^2+2UV+V^2-W^2 = (U+V+W)(U+V-W),$$

$$U^3+V^3+W^3-3UVW = (U+V+W)(U^2+V^2+W^2-UV-UW-VW),$$

$$U^2-4TUVW-T^2V^4-4T^2W^4 = (U-2TVW-TV^2-2TW^2)(U-2TVW+TV^2+2TW^2),$$

where $T,U,V,W$ are monomials in $x_1,\ldots,x_\ell$ (see Fried and Schinzel 1972).

## Section 15. Gourin's theorem

In this section let $k$ be an algebraically closed field and suppose a polynomial $F(x_1,\ldots,x_s)$ to be irreducible over $k$ and contain more than two terms.

Theorem 23. (Gourin 1931, for $k=\mathbb{C}$). *Let* $\langle t_1,\ldots,t_s\rangle$ *be an integral vector, where* $t_i>0$, $t_i \not\equiv 0 \pmod{\mathrm{char}\ k}$. *Then there exists an integral vector* $\langle \tau_1,\ldots,\tau_s\rangle$ *such that*

(i) $\quad 0 < \tau_i \leq |F|^2$,

(ii) $\quad t_i = \tau_i u_i, \quad u_i \in Z$

(iii) $\quad F(x_1^{\tau_1},\ldots,x_s^{\tau_s}) \underset{\bar{k}}{\overset{can}{=}} \mathrm{const} \prod_{\rho=1}^{r} F_\rho(x_1,\ldots,x_s)^{e_\rho}$

*implies* $e_\rho=1$ *and*

$$F(x_1^{t_1},\ldots,x_s^{t_s}) \underset{\bar{k}}{\overset{can}{=}} \mathrm{const} \prod_{\rho=1}^{r} F_\rho(x_1^{u_1},\ldots,x_s^{u_s}).$$

We recall that $|F| = \max_{1\leq i\leq s} |F|_{x_i}$.

Remark. The folowing example shows that the bound given in (i) is best possible. Consider the expression

$$F(x_1,\ldots,x_s) = \prod_{j=0}^{m^2-1} (1+\zeta_{m^2}^j\, x_1^{1/m^2} x_2^{1/m} \ldots x_s^{1/m} + \zeta_m^j\, x_1^{1/m}).$$

It is unchanged under substitutions $x_1^{1/m^2} \to \zeta_{m^2}^{h_1} x_1^{1/m^2},\ldots,$ $x_i^{1/m} \to \zeta_m^{h_i} x_i^{1/m}$ $(2\leq i\leq s)$.

Indeed, $F \longrightarrow \prod_{j=0}^{m^2-1} \left(1 + \zeta_{m^2}^{j+h_1+m(h_2+\ldots+h_s)} x_1^{1/m^2} x_2^{1/m} \ldots x_s^{1/m} + \zeta_m^{j+h_1} x_1^{1/m}\right) = F.$

Hence $F$ is a polynomial and $|F| = m$. Take $m \not\equiv 0 \pmod{\text{char } k}$, $\langle t_1,\ldots,t_s\rangle = \langle m^2, m,\ldots,m\rangle$. Clearly $F(x_1^{t_1},\ldots,x_s^{t_s}) = \prod_{j=0}^{m^2-1}(1+\zeta_{m^2}^j\, x_1\, x_2 \ldots x_s + \zeta_m^j\, x_1^m)$. Values of $u_1,\ldots,u_s$ must be 1, $\tau_i = t_i$. Thus $\tau_1 = t_1 = |F|^2$.

Note that the preceding $F$ is irreducible. If $F = AB$, then $F(x_1^{t_1},\ldots,x_s^{t_s}) = A(x_1^{m^2}, x_2^m,\ldots,x_s^m)\, B(x_1^{m^2}, x_2^m,\ldots,x_s^m)$ and $1+\zeta_{m^2}\, x_1 \ldots x_s + \zeta_m x_1^m \mid A(x_1^{m^2}, x_2^m,\ldots,x_s^m)$ implies $B = \text{const}$.

Convention. Two polynomials are equivalent if they differ by a constant factor $\neq 0$. A polynomial $F$ is primary in $x_i$ if the exponents of $x_i$ in $F$ are coprime. $F$ is primary if $F$ is primary with respect to all its variables.

Lemma 1. If $F(x_1^{t_1},\ldots,x_s^{t_s}) \underset{k}{\overset{\text{can}}{=}} \text{const} \prod_{\rho=1}^{N} \tilde{F}_\rho(x_1,\ldots,x_s)^{\tilde{e}_\rho}$, then $\tilde{F}_\rho$ form a complete set of nonequivalent polynomials derived from $\tilde{F}_1$ by substitutions of the form $x_1 \longrightarrow \zeta_{t_1}^{i_1} x_1,\ldots,x_s \longrightarrow \zeta_{t_s}^{i_s} x_s$ and $\tilde{e}_\rho$ are all equal to 1.

Proof. Clearly all polynomials derived from $\tilde{F}_1$ are irreducible and if nonequivalent are pairwise coprime. Hence $F(x_1^{t_1},\ldots,x_s^{t_s}) = PR$, where $P$ is the product of all inequivalent polynomials derived from $\tilde{F}_1$. Every substitution (*) $x_1 \longrightarrow \zeta_{t_1}^{i_1} x_1$ transforms $P$ into an equivalent polynomial. If $x_1 \mid P$ then $x_1 \mid F$ and either $F$ is reducible or it has just one term. Hence $P$ contains some term not depending on $x_1$, thus it is invariant under

all substitutions of the type (*) and $P = P_1(x_1^{t_1}, x_2, .., x_s)$.
By symmetry we have $P = P_o(x_1^{t_1}, \dots, x_s^{t_s})$, hence

$$R = \frac{F(x_1^{t_1}, \dots, x_s^{t_s})}{P(x_1^{t_1}, \dots, x_s^{t_s})} \in k\,[x_1^{t_1}, \dots, x_s^{t_s}] \text{ and we have}$$

$F = P_o R_o$, contradicting assumption that $F$ is irreducible in $k$, unless $R = R_o =$ const.

Corollary. All $\tilde{F}_\rho$ contain all variables.

Lemma 2. Let $\tilde{F}_1 = \sum_{j=0}^{h} a_j \prod_{i=1}^{s} x_i^{\alpha_{ij}}$, where $\alpha_{10} = 0$, $a_j \neq 0$, $\langle \alpha_{1j}, \dots, \alpha_{sj} \rangle$ are distinct, and $h \geq 2$. If $\tilde{F}_1$ is primary in $x_1$ and irreducible, then for at least one i the vectors $\langle \alpha_{11}, \dots, \alpha_{1h} \rangle$ and $\langle \alpha_{i1} - \alpha_{io}, \dots, \alpha_{ih} - \alpha_{io} \rangle$ are linearly independent.

Proof. By the assumption $(\alpha_{11}, \dots, \alpha_{1h}) = 1$.

If $\langle \alpha_{i1} - \alpha_{io}, \dots, \alpha_{ih} - \alpha_{io} \rangle = \beta_i \langle \alpha_{11}, \dots \alpha_{1h} \rangle$, where for all i we have $\beta_i \in \mathbf{Z}$, then

$$\tilde{F}_1 \prod_{i=1}^{s} x_i^{-\alpha_{io}} = a_o + \sum_{j=1}^{h} a_j \left( \prod_{i=1}^{s} x_i^{\beta_i} \right)^{\alpha_{1j}} = G\left( \prod_{i=1}^{s} x_i^{\beta_i} \right).$$

Since $G$ has more than two terms and $k$ is algebraically closed we have $G = G_1 G_2$, $G_\nu \in k[x] \setminus k\ (\nu = 1, 2)$.

Hence

$$G\left( \prod_{i=1}^{s} x_i^{\beta_i} \right) = G_1\left( \prod_{i=1}^{s} x_i^{\beta_i} \right) \ G_2\left( \prod_{i=1}^{s} x_i^{\beta_i} \right)$$

and

$$\tilde{F}_1(x_1, \dots, x_s) = JG_1\left( \prod_{i=1}^{s} x_i^{\beta_i} \right) \ JG_2\left( \prod_{i=1}^{s} x_i^{\beta_i} \right),$$

contradicting irreducibility of $\tilde{F}_1$.

<u>Lemma 3.</u> If under the assumptions of Theorem 23,

$$F(x_1^{t_1},\ldots,x_s^{t_s}) \underset{\bar{k}}{\overset{can}{=}} \prod_{\rho=1}^{N} \tilde{F}_\rho(x_1,\ldots,x_s), \text{ and } \tilde{F}_1 \text{ satisfies}$$

the condition of Lemma 2, then $t_1 \leq |F|^2$.

<u>Proof</u>. By Lemma 2, we can assume the minor

$$D = \begin{vmatrix} \alpha_{11} & \alpha_{12} \\ \alpha_{21}-\alpha_{20} & \alpha_{22}-\alpha_{20} \end{vmatrix}$$

to be positive.

Let $|\tilde{F}_1|_{x_i} = n_i$, $|F|_{x_i} = m_i$, $i = 1,2$. We have $D \leq n_1 n_2$. For, since $D = \alpha_{11}(\alpha_{22}-\alpha_{20})-\alpha_{12}(\alpha_{21}-\alpha_{20})$ and $n_i \geq \alpha_{ij} \geq 0$ we have

$$|D| \leq n_1 n_2 \quad \text{if} \quad (\alpha_{22}-\alpha_{20})(\alpha_{21}-\alpha_{20}) \geq 0$$

and $|D| \leq n_1 |\alpha_{22}-\alpha_{20}-\alpha_{21}+\alpha_{20}| \leq n_1 n_2$

$$\text{if} \quad (\alpha_{22}-\alpha_{20})(\alpha_{21}-\alpha_{20}) < 0.$$

If we make the substitution $x_1 \longrightarrow \zeta_{t_1}^d x_1$, $(0 \leq d \leq t_1 - 1)$ then claim

(1) $\tilde{F}_1(\zeta_{t_1}^d x_1, x_2, \ldots, x_s) \neq \text{const}\, \tilde{F}_1(x_1, x_2, \ldots, x_s)$ if $d \neq 0$.

Suppose otherwise. Since $\alpha_{10} = 0$ there is a term independent of $x_1$ in $\tilde{F}_1$, hence $a_j \zeta_{t_1}^{d\alpha_{1j}} = a_j$, $d\alpha_{1j} \equiv 0 \pmod{t_1}$,

$j \leq h$. Since $\tilde{F}_1$ is primary in $x_1$ it follows that $d \equiv 0$ (mod $t_1$), a contradiction.

(1) implies, in virtue of Lemma 1, that the degree in $x_1$ of $\prod_{\rho=1}^{N} \tilde{F}_\rho (x_1,\ldots,x_s)$ is at least $n_1 t_1$.

Since that of $F(x_1^{t_1},\ldots,x_s^{t_s})$ is $m_1 t_1$ we get $n_1 t_1 \leq m_1 t_1$, $n_1 \leq m_1$. Now replace $x_2$ by $\zeta_{t_2}^{e} x_2$. If $\tilde{F}_1(\zeta_{t_1}^{d} x_1, \zeta_{t_2}^{e} x_2, x_3, \ldots, x_s)$ $= \text{const}\ \tilde{F}_1(x_1,\ldots,x_s)$ we get $a_j \zeta_{t_1}^{d\alpha_{1j}} \zeta_{t_2}^{e\alpha_{2j}} = \text{const}\ a_j$ (for all j)

Putting $d = 0$ we get $\text{const} = \zeta_{t_1}^{d\alpha_{10}} \zeta_{t_2}^{e\alpha_{20}} a_j = \zeta_{t_2}^{e\alpha_{20}} a_j, (\alpha_{10} = 0)$, hence

$$\zeta_{t_1}^{d\alpha_{1j}} \zeta_{t_2}^{e(\alpha_{2j} - \alpha_{20})} = 1 \ (1 \leq j \leq h).$$

The equations

$$\zeta_{t_1}^{d\alpha_{11}} \zeta_{t_2}^{e(\alpha_{21} - \alpha_{20})} = 1 \quad \text{and}$$

$$\zeta_{t_1}^{d\alpha_{12}} \zeta_{t_2}^{e(\alpha_{22} - \alpha_{20})} = 1$$

imply $\zeta_{t_2}^{eD} = 1$. Hence

$$eD \equiv 0 \pmod{t_2},$$

$$e \equiv 0 \pmod{\frac{t_2}{(t_2,D)}}.$$

Thus if $e = 0,\ldots,\dfrac{t_2}{(t_2,D)} - 1$ the polynomials obtained from $\tilde{F}_1$ will be inequivalent. If, as before, we compare the degrees we get

$$n_2 t_1 \frac{t_2}{(t_2,D)} \le t_2 m_2, \quad n_2 t_1 \le m_2 (t_2,D) \le m_2 n_1 n_2,$$

hence

$$t_1 \le n_1 m_2 \le m_1 m_2 \le |F|^2.$$

Lemma 4. If under the assumptions of Lemma 3 $F$ has more than two terms and $\tilde{F}_1$ is primary then $\tilde{F}_1$ has more than two terms.

Proof. If not, then by Corollary to Lemma 1 after a suitable **renumbering** of variables

$$\tilde{F}_1 = a_1 x_1 \dots x_\ell + b x_{\ell+1} \dots x_s$$

$$= a x_1 x_2 \dots x_\ell (1 + cq), \qquad q = \frac{x_{\ell+1} \dots x_s}{x_1 \dots x_\ell} .$$

Lemma 1 implies that $F(x_1^{t_1}, \dots, x_s^{t_s}) = ca^N (x_1 \dots x_\ell)^N (1+c_1 q) \dots$

$\dots (1+c_N q) = ca^N (x_1 \dots x_\ell)^N (a + e_{\nu_1} q^{\nu_1} + \dots + e_N q^N), \quad e_{\nu_i} \ne 0.$

The left hand side is invariant under $x_i \longrightarrow x_i \zeta_{t_i}^{d_i}$, so

$N \equiv 0 (\mathrm{mod}\ T)$, where $T = [t_1, \dots, t_s]$, and $\nu_i \equiv 0 (\mathrm{mod}\ T)$.

Since $F$ has more than two terms we have $N > T$ and

$F(x_1^{t_1}, \dots, x_s^{t_s}) = ca^N (x_1 \dots x_\ell)^N \ G_1(q^T) G_2(q^T)$, where

$|G_1| \ge 1, \quad |G_2| \ge 1.$

Let $\frac{N}{t_i} = \lambda_i \quad (1 \le i \le s), \quad g_\nu (x_1, \dots, x_s) =$

$= G_\nu \left( \frac{x_{\ell+1}^{\lambda_{\ell+1}} \dots x_s^{\lambda_s}}{x_1^{\lambda_1} \dots x_s^{\lambda_s}} \right) \qquad (\nu = 1,2).$ Then

$$F(x_1^{t_1},\ldots,x^{t_s}) = ca^N(x_1^{t_1})^{\lambda_1}\ldots(x_\ell^{t_\ell})^{\lambda_s}\, g_1(x_1^{t_1},\ldots,x_s^{t_s})g_2(x_1^{t_1},\ldots$$

$$\ldots,x_s^{t_s}) \quad \text{and} \quad G(x_1,\ldots,x_s) = Jg_1(x_1,\ldots,x_s)Jg_2(x_1,\ldots,x_s),$$

contradicting irreducibility of $F$. Thus $\tilde{F}_1$ has more than two terms.

Proof of Theorem 23.

Suppose first that $F$ is primary and let

$$F(x_1^{t_1},\ldots,x_s^{t_s}) \underset{k}{\overset{can}{=}} \text{const} \prod_{\rho=1}^{N} \tilde{F}_\rho(x_1,\ldots,x_s)^{\tilde{e}_\rho},$$

$\tilde{F}_1(x_1,\ldots,x_s) = F_1(x_1^{u_1},\ldots,x_s^{u_s})$ where $F_1$ is primary.

By Lemma 1 we have $\tilde{F}_\rho = F_\rho(x_1^{u_1},\ldots,x_s^{u_s})$ for a suitable primary $F_\rho$, $\tilde{e}_\rho = 1$ and

$$t_i = u_i\tau_i, \quad \tau_i \in \mathbb{Z} \quad (\text{since } F \text{ is primary}),$$

$$F(x_1^{\tau_1},\ldots,x_s^{\tau_s}) \underset{k}{\overset{can}{=}} \prod_{\rho=1}^{N} F_\rho(x_1,\ldots,x_s).$$

By Lemma 4 $F_1$ has more than two terms.

By Lemma 3 $\tau_i \leq |F|^2$ $(1 \leq i \leq s)$.

If $F$ is not primary then $F = P(x_1^{\nu_1},\ldots,x_s^{\nu_s})$ where $P$ is primary. Let $\nu_i = \pi_i\rho_i$, where $\pi_i \geq 1$ is a power of char $k$ and $\rho_i \not\equiv 0$ mod char $k$. We apply the previous case to $\langle\rho_1t_1,\ldots,\rho_st_s\rangle$ and infer the existence of $\tau_1',\ldots,\tau_s'$ such that all $i \leq s$ we have $\rho_it_i = \tau_i'u_i'$, $u_i' \in \mathbb{Z}$, $\tau_i' \leq |P|^2$ and if

$$P(x_1^{\tau_1'},\ldots,x_s^{\tau_s'}) \underset{k}{\overset{can}{=\!=}} \text{const} \prod_{\rho=1}^{r} P_\rho(x_1,\ldots,x_s)^{e_\rho},$$

then $P_\rho$ is primary, $e_\rho = 1$ and

$$P(x_1^{\rho_1 t_1},\ldots,x_s^{\rho_s t_s}) \underset{k}{\overset{can}{=\!=}} \text{const} \prod_{\rho=1}^{r} P_\rho(x_1^{u_1'},\ldots,x_s^{u_s'}).$$

Define

$$\tau_i = \frac{[\rho_i,\tau_i']}{\rho_i} = \frac{\tau_i'}{(\rho_i,\tau_i')} \le |P|^2 \le |F|^2, \quad u_i = \frac{t_i}{\tau_i} = (u_i',t_i)$$

$$F_\rho(x_1,\ldots,x_s) = P_\rho(x_1^{\pi_1[\rho_1,\tau_1']/\tau_1'},\ldots,x_s^{\pi_s[\rho_s,\tau_s']/\tau_s'}).$$

We have

$$F(x_1^{\tau_1},\ldots,x_s^{\tau_s}) = P(x_1^{\pi_1 2\rho_1\tau_1},\ldots,x_s^{\pi_s\rho_s\tau_s}) = P(x_1^{\pi_1[\rho_1,\tau_1']},\ldots$$

$$\ldots,x_s^{\pi_s[\rho_s,\tau_s']}) = \text{const} \prod_{\rho=1}^{r} F_\rho(x_1,\ldots,x_s)$$

and $F(x_1^{t_1},\ldots,x_s^{t_s}) = \text{const} \prod_{\rho=1}^{r} F_\rho(x_1^{u_1},\ldots,x_s^{u_s}).$

Since $F$ is irreducible in $k$ we have $\pi_i = 1$ for a certain $i \le s$. But then since $P_\rho$ is primary in $x_i$ and $u_i' \not\equiv 0 \bmod \text{char}\, k$, in virtue of Corollary 2 to Theorem 21 irreducibility of $P_\rho(x_1^{u_1'},\ldots,x_s^{u_s'})$ implies

irreducibility of $P_\rho(x_1^{\pi_1 u_1'},\ldots,x_s^{\pi_s u_s'}) = F_\rho(x_1^{u_1},\ldots,x_s^{u_s})$

and a fortiori of $F_\rho(x_1,\ldots,x_s)$. It remains to

show that for different $\rho$ polynomials $F_\rho(x_1^{u_1},\ldots,x_s^{u_s})$ are coprime. This follows from the corresponding property of $P_\rho$ and Lemma 1 of the next section.

Remark. Without the assumption $t_i \not\equiv 0 \bmod \operatorname{char} k$ for all $i$ Theorem 23 is not true as it stands. It can however be suitably modified provided

$(t_1,\ldots,t_s) \not\equiv 0 \bmod \operatorname{char} k$.

## Section 16. Extension of Gourin's theorem to arbitrary fields

Theorem 24. (Schinzel 1978 for $k = \mathbb{Q}$). *Let* $F(\mathbf{x}) \neq \text{const}\, x_i$ *be irreducible over* $k$ *and different from* $JF_0(x_1^{\delta_1} \cdots x_s^{\delta_s})$ *for any* $F_0$ *and any nonzero integral vector* $\langle \delta_1, \ldots, \delta_s \rangle$. *For any integral vector* $\langle t_1, \ldots, t_s \rangle$, *where* $t_i > 0$ *and* $t_i \not\equiv 0 \pmod{\text{char}\, k}$ *there exists an integral vector* $\langle \tau_1, \ldots, \tau_s \rangle$ *such that* $0 < \tau_i \leq |F|^2$, $t_i = \tau_i u_i$, *and if*

$$F(x_1^{\tau_1}, \ldots, x_s^{\tau_s}) \underset{k}{\overset{\text{can}}{=}} \text{const} \prod_{\rho=1}^{r} F_\rho(x_1, \ldots, x_s)^{e_\rho} \quad \text{then} \quad e_\rho = 1$$

*and* $F(x_1^{t_1}, \ldots, x_s^{t_s}) \underset{k}{\overset{\text{can}}{=}} \text{const} \prod_{\rho=1}^{r} F_\rho(x_1^{u_1}, \ldots, x_s^{u_s})$.

Lemma 1. If $(P(x_1,\ldots,x_s), Q(x_1,\ldots,x_s)) = G(x_1,\ldots,x_s)$ then $(P(x_1^{t_1},\ldots,x_s^{t_s}), Q(x_1^{t_1},\ldots,x_s^{t_s})) = G(x_1^{t_1},\ldots,x_s^{t_s})$.

Proof. Clearly $(P_0,Q_0) = 1$, where $P_0 = P/G$, $Q_0 = Q/G$. Take the resultant $R$ of $P_0,Q_0$ with respect to $x_1$. There exist polynomials $U,V$ such that

$$P_0U + Q_0V = R.$$

Put in powers of $x_i$. It follows that

$$(P_0(x_1^{t_1},\ldots,x_s^{t_s}), \quad Q_0(x_1^{t_1},\ldots,x_s^{t_s}))$$

is independent of $x_1$ and by symmetry of any $x_i$. Hence the result.

Lemma 2. Let $\Psi(\mathbf{x})$ be a monic polynomial with coefficients algebraic over $k$ and $\Omega$ the field gotten by adjoining the coefficients of $\Psi$ to $k$. If $\Psi$ is irreducible over the normal closure of $\Omega$ over $k$ then $N_{\Omega/k}\Psi(\mathbf{x})$ is irreducible over $k$.

Proof. Let $\Omega_o$ be the maximal separable extension of $k$ in $\Omega$. Then $\Omega/\Omega_o$ is purely inseparable of degree $d \geq 1$, where $d$ is the least exponent such that $\Psi^d = \Phi_o \in \Omega_o[\mathbf{x}]$. Suppose that $\Phi_1$ is a monic factor of $\Phi_o$ in the normal closure $N_o$ of $\Omega_o$ over $k$. Then $\Phi_1 = \Psi^e \in N_o[\mathbf{x}]$ and since $\Phi^d \in N_o[\mathbf{x}]$ we get $\Psi^{(d,e)} \in N_o[\mathbf{x}]$. An arbitrary coefficient of $\Psi^{(d,e)}$ is separable over $\Omega_o$ and belongs to $\Omega$ thus it belongs to $\Omega_o$. Hence $\Psi^{(d,e)} \in \Omega_o[\mathbf{x}]$, $(d,e) = d$ and $e = d$. Thus $\Psi^d$ is irreducible over the normal closure of $\Omega_o$ over $k$. But

$$N_{\Omega/k}(\Psi(\mathbf{x})) = N_{\Omega_o/k}(N_{\Omega/\Omega_o}\Psi(\mathbf{x})) = N_{\Omega_o/k}\Psi(\mathbf{x})^d,$$

so it suffices to prove the lemma in the case, where $\Omega/k$ is separable. Then $N_{\Omega/k}\Psi = \Pi_\sigma \Psi^\sigma$ where the product is taken over all isomorphic injections $\sigma$ of $\Omega$ into its algebraic closure stable on $k$. But by the definition of $\Omega$ we have $\Psi^\sigma \neq \Psi$ for any $\sigma$ different from identity. In fact, since $\Psi$ is monic, $\Psi^\sigma \neq \text{const}\,\Psi$. The polynomials $\Psi^\sigma$ are relatively prime, and if $\Psi|\Phi$, $\Phi$ irreducible over $k$, and monic, then

$$\Pi_\sigma \Psi^\sigma | \Phi .$$

Hence $N_{\Omega/k}\Psi|\Phi$ and $N_{\Omega/k}\Psi = \Phi$ is irreducible over $k$.

Remark. If $\Psi$ is irreducible merely over $\Omega$ and not over its normal closure the conclusion need not hold as is shown by the example $\Psi(x) = x^2 + \sqrt[3]{2}\,x + \sqrt[3]{4}$.

Proof of Theorem 24. Let $\Psi|F(x_1,\ldots,x_s)$, $\Psi$ irreducible over the algebraic closure $\hat{k}$ of $k$ and monic. Let the coefficients of $\Psi$ generate $\Omega$. By Corollary 1 to Theorem 10, $[\Omega : k] < \infty$. By Lemma 2, $N_{\Omega/k}\Psi$ is irreducible over k, hence $F = \text{const } N_{\Omega/k}\Psi$. By Theorem 23, there exist $\tau_i$ auch that $|\tau_i| \leq |\Psi|^2$, $t_i = \tau_i u_i$ $(1 \leq i \leq s)$ and every irreducible factor of $\Psi(x_1^{t_1},\ldots,x_s^{t_s})$ in $\hat{k}$ is of form $T(x_1^{u_1},\ldots,x_s^{u_s})$. Take these $\tau_i$. For the last condition of Theorem 24 we know by Lemma 1 that $F_\rho(x_1^{u_1},\ldots,x_s^{u_s})$ are prime to each other. We need to show that they are irreducible. Let $\Psi_\rho$ be a factor of $F_\rho(x_1^{u_1},\ldots,x_s^{u_s})$, irreducible over $\hat{k}$ and monic. Coefficients of $\Psi_\rho$ generate $\Omega_\rho$, $[\Omega_\rho : k] < \infty$. Since $\Psi_\rho | F(x_1^{t_1},\ldots,x_s^{t_s}) = \text{const } N_{\Omega/k}\Psi(x_1^{t_1},\ldots,x_s^{t_s})$ we can assume without loss of generality that $\Psi_\rho | \Psi(x_1^{t_1},\ldots,x_s^{t_s})$. Hence $\Psi_\rho = T(x_1^{u_1},\ldots,x_s^{u_s})$. By Lemma 2, $N_{\Omega/k}\Psi_\rho =$ $= N(x_1^{u_1},\ldots,x_s^{u_s})$ is irreducible over k. Hence $N(x_1^{u_1},\ldots,x_s^{u_s}) | F_\rho(x_1^{u_1},\ldots,x_s^{u_s})$, $N | F_\rho$ and since $F_\rho$ is irreducible, $N = \text{const } F_\rho$.

Thus $F_\rho(x_1^{u_1},\ldots,x_s^{u_s}) = \text{const } N(x_1^{u_1},\ldots,x_s^{u_s})$ is irreducible over k.

It remains to show that all the exponents $e_\rho$ are equal to 1. To do this we first factorize F over the field $\tilde{k} = k\ (\zeta_{t_1},\ldots,\zeta_{t_s})$. Let $\tilde{F}$ be a monic irreducible factor of F over $\tilde{k}$ and $\Omega$ be the field generated over k by the coefficients of $\tilde{F}$. Since $\tilde{k}/k$ is normal and separable $N_{\Omega/k}\tilde{F}$ is irreducible over k by Lemma 2 and we have

$$F = \text{const } N_{\Omega/k}\tilde{F} = \text{const } \prod_{\sigma} \tilde{F}^{\sigma}$$

where $\sigma$ runs through all isomorphic injections of $\Omega$ into $\tilde{k}$ stable on $k$. Furthermore the factors on the right side are relatively prime in pairs. Now, since $\tilde{k}$ contains $\zeta_{t_i}$ for all $i \leq s$ the argument leading to Lemma 1 of Theorem 23 applies and by that lemma $\tilde{F}^{\sigma}(x_1^{t_1},\ldots,x_s^{t_s})$ has no multiple factors over $\tilde{k}$. Since $(\tilde{F}^{\sigma_1}, \tilde{F}^{\sigma_2}) = 1$ implies by Lemma 1 that

$$(\tilde{F}^{\sigma_1}(x_1^{t_1},\ldots,x_s^{t_s}), \tilde{F}^{\sigma_2}(x_1^{t_1},\ldots,x_s^{t_s})) = 1,$$

it follows that $F(x_1^{t_1},\ldots,x_s^{t_s})$ has no multiple factors over $\tilde{k}$ and a fortiori over $k$.

## Section 17. Kneser's theorem and an extension of Capelli's theorem

Theorem 25 (M.Kneser 1975). *Let $K/k$ be a separable extension, and let $M$ be a subgroup of $K^*$ such that $[Mk^* : k^*] < \infty$. Then*

$$(*) \qquad [k(M) : k] = [k^*M : k^*]$$

*if and only if (i) for all primes $p$, $\zeta_p \in k^*M$ implies $\zeta_p \in k$ and (ii) $1 + \zeta_4 \in k^*M$ implies $\zeta_4 \in k$.*

Proof $(*) \longrightarrow$ (i) & (ii). We clearly have

$$[k(M) : k] \leq [k^*M : k^*],$$

$$(1) \qquad [k(M) : k] = [k(M) : k(\zeta_p)][k(\zeta_p) : k]$$

and

$$(2) \qquad [k^*M : k^*] = [k^*M : k^*<\zeta_p>][k^*<\zeta_p> : k^*].$$

The factors on the right hand side of (1) do not exceed the corresponding factors on the right hand side of (2). If the left hand sides are equal then $[k^*<\zeta_p> : k^*] = [k(\zeta_p) : k]$, thus $\zeta_p \in k$. In the above argument we can replace $\zeta_p$ by $1 + \zeta_4$.

(i) & (ii) $\longrightarrow$ $(*)$. It is enough to prove the theorem when $[k^*M : k^*]$ is of prime power order. Then if in the general case $p^\nu \| [k^*M : k^*]$ we take the Sylow subgroup $S_{p^\nu}$ of $k^*M/k^*$ and the corresponding union of cosets mod $k^*$ that we denote by $k^*S_{p^\nu}$. If for all such subgroups

$[k^*S_{p^\nu} : k^*] = [k(k^*S_{p^\nu}) : k]$ we have $[k(M) : k] \equiv 0 \bmod p^\nu$ for all $p$ and

$$[k(M) : k] = [k^*M : k^*].$$

Therefore, assume that $[k^*M : k^*] = p^\nu$ and let

$$k^* = N_o \subset N_1 \subset N_2 \ldots \subset N_\nu = k^*M,$$

where $[N_s : N_{s-1}] = p$.

We will prove by induction on $s$ the following two statements:

$(A_s)$ $[k(N_s) : k(N_{s-1})] = p$.

$(B_s)$ If $p > 2$ and $c \in k(N_s)$ and $c^p \in N_s$, then $c \in N_s$.

If $p = 2$ and $c \in k(N_s)$ and $c^2 \in N_s$, and either $c \in k^*M$ or $\zeta_4 \notin k(N_s)$, then $c \in N_s$.

$B_o$ is trivially true and $A_o$ has no meaning.

We shall show that $B_{s-1}$ implies $A_s$. Let $N_s = N_{s-1}\langle a\rangle$, where $a \notin N_{s-1}$. Clearly over $k(N_{s-1})$ $a$ satisfies the equation $x^p - a^p = 0$. If this is reducible over $k(N_{s-1})$ then by Lemma 1 to Theorem 21 $a^p = b^p$ for some $b \in k(N_{s-1})$, $b^p \in N_{s-1}$. If $p > 2$, by $B_{s-1}$ on taking $b = c$ it follows that $b \in N_{s-1}$. Since $a = b\zeta_p$, $a \in N_s$, $b \in N_{s-1}$ we have $\zeta_p \in N_s \subset k^*M$. We infer from the condition (i) that $\zeta_p \in k$ thus $a \in N_{s-1}$, contrary to the choice of a. If $p = 2$, $a^2 = b^2$ we have $a = \pm b$ and $b \in N_s \subset k^*M$. Taking $b = c$ in $B_{s-1}$ we infer as before that $b \in N_{s-1}$ thus again $a \in N_{s-1}$. Thus for all $p$ we must have that $x^p - a^p$ is irreducible over $k(N_{s-1})$ and $A_s$ holds. We next proceed to show that $B_{s-1}$ and $A_s$ together imply $B_s$.

Let $p > 2$, $c \in k(N_s)$ and $c^p \in N_s$. Then $c^p = a^q d$, where $d \in N_{s-1}$, $0 \leq q < p$ and $N_s = N_{s-1}\langle a \rangle$, $a \notin N_{s-1}$. Assume first $q > 0$. Taking the norm $N$ from $k(N_s)$ to $k(N_{s-1})$ we get

$$Nc^p = Na^q Nd,$$

$$(Nc)^p = (Na)^q d^p.$$

Since by $A_s$ the minimal polynomial of a over $k(N_{s-1})$ is $x^p - a^p$ we have $Na = a^p$. Thus

$$(Na)^q d^p = a^{pq} d^p \quad \text{and}$$

$$\left(\frac{Nc}{d}\right)^p = (a^p)^q.$$

Since $(q,p) = 1$ it follows that

$$a^p = f^p \quad \text{for some} \quad f \in k(N_{s-1})$$

Since $a^p \in N_{s-1}$ we have $f^p \in N_{s-1}$ and on applying $B_{s-1}$ we get $f \in N_{s-1}$. Now $a = \zeta_p f$, hence $\zeta_p \in N_s \subset k^*M$ and by the condition (i) $\zeta_p \in k$, $a \in N_{s-1}$, a contradiction. Thus $q = 0$ and $c^p = d \in N_{s-1}$. Take an isomorphism $S$ of the normal closure of $k(N_s)$ stable on $k(N_{s-1})$ and such that $Sa = a\zeta_p$. We have

$$(Sc)^p = c^p, \quad \text{hence}$$

$$Sc = c\,\zeta_p^j, \quad S(ca^{-j}) = c\,\zeta_p^j\,a^{-j}\zeta_p^{-j} = ca^{-j}.$$

It follows that $ca^{-j} \in k(N_{s-1})$. Since $c^p a^{-jp} \in N_{s-1}$ the statement $B_{s-1}$ implies $ca^{-j} \in N_{s-1}$. Hence $c \in N_s$.

Now take $p = 2$. We have either $c^2 = ad$ or $c^2 = d$, $d \in N_{s-1}$. If

$$c^2 = ad \tag{3}$$

we get $(Nc)^2 = Na \cdot d^2 = -a^2d^2$, whence $\pm Nc = f = \zeta_4 ad$.

Since $f \in k(N_{s-1})$, $d \in k(N_{s-1})$, and $a \notin k(N_{s-1})$ it follows that $\zeta_4 \notin k(N_{s-1})$, but $\zeta_4 \in k(N_s)$. Thus $\zeta_4$ generates $k(N_s)/k(N_{s-1})$ and

$$c = g + \zeta_4 h, \quad \text{where} \quad g,h \in k(N_{s-1}).$$

Squaring we get $c^2 = g^2 - h^2 + 2gh\,\zeta_4$

and since by (3) $c^2 = -\zeta_4 f$

it follows that $g^2 = h^2$, $g = \pm h$, whence

$$c = g(1 \pm \zeta_4). \tag{4}$$

But then $c^4 = -4g^4$ and since $c^2 \in N_s$, $g^4 = -\frac{c^4}{4} \in N_{s-1}$.

By $B_{s-1}$, $g^2 \in N_{s-1}$; applying $B_{s-1}$ again we get $g \in N_{s-1}$. If $c \in k^*M$ (4) gives $1 \pm \zeta_4 \in k^*M$ and by the condition (ii) of the theorem $\zeta_4 \in k^*$. Hence $c \in N_{s-1}$.

It remains to consider the case, where

$$c^2 = d \in N_{s-1}.$$

We use the same trick as for $p > 2$, taking an isomorphism $S$ of $k(N_s)$ stable on $k(N_{s-1})$ and such that $Sa = -a$. We have for a suitable $j$

$$S(ca^j) = ca^j, \quad ca^j \in k(N_{s-1}).$$

Since $c^2a^{2j} \in N_{s-1}$ the statement $B_{s-1}$ implies $ca^j \in N_{s-1}$, thus $c \in N_s$.
The inductive proof of $A_s$ is thus complete and the theorem follows.

Corollary 1. (Hasse 1930). Let $k$ be an algebraic number field, containing $\zeta_n$. Let $\xi_1,\dots,\xi_\ell$ be algebraic numbers and assume $\xi_i^n \in k$ $(1 \le i \le \ell)$, and

$$\xi_1^{x_1}\dots\xi_\ell^{x_\ell} \in k \text{ implies } \xi_i^{x_i} \in k \quad (1 \le i \le \ell). \tag{5}$$

Then we have

$$[k(\xi_1,\dots,\xi_\ell) : k] = \prod_{i=1}^{\ell} [k^*\langle\xi_i\rangle : k^*].$$

Proof. In Theorem 25 take $M = k^*\langle\xi_1,\ldots,\xi_\ell\rangle$. Since $M^n \subseteq k^*$, $\zeta_p \in k^*M$ implies $\zeta_p^n \in k$ and hence $\zeta_p \in k$ if $p \nmid n$. If $p \mid n$ the same holds by the assumption $\zeta_n \in k$. Thus the condition (i) is satisfied. Similarly $1 + \zeta_4 \in k^*M$ implies $(1 + \zeta_4)^n \in k$, hence $\zeta_4 \in k$ if $4 \nmid n$. Thus the condition (ii) is satisfied. By the theorem

$$[k(\xi_1,\ldots,\xi_\ell) : k] = [k^*\langle\xi_1,\ldots,\xi_\ell\rangle : k^*]$$

and the right hand side equals by virtue of (5)

$$\prod_{i=1}^{\ell} [k^*\langle\xi_i\rangle : k^*].$$

Corollary 2. (Mordell 1955). Let $k \subset \mathbb{R}$, $\xi_i \in \mathbb{R}$; $\xi_i^{n_i} \in k$ $(1 \le i \le \ell)$. If $\xi_1^{x_1}\cdots\xi_\ell^{x_\ell} \in k$ implies $\xi_i^{x_i} \in k$ for all $i \le \ell$ then

$$[k(\xi_1,\ldots,\xi_\ell) : k] = [k^*\langle\xi_1,\ldots,\xi_\ell\rangle : k^*] = \prod_{i=1}^{\ell} [k^*\langle\xi_i\rangle : k^*].$$

Corollary 3. (Siegel 1972). Let $k \subset \mathbb{C}$, $\xi_i^{n_i} \in k$ $(1 \le i \le \ell)$. If either $\zeta_{n_i} \in k$ or $k$ and $\xi_i$ are real, then

$$[k(\xi_1,\ldots,\xi_\ell) : k] = [k^*\langle\xi_1,\ldots,\xi_\ell\rangle : k^*].$$

Corollary 3 implies Corollary 2 and is itself deduced from the theorem in the same way as Corollary 1.

Corollary 4. (Schinzel 1975 b). Let $p$ be a prime different from char. $k$, $\zeta_p \in k$, $\xi^{p^\mu} \in k^*$, $\eta^{p^\nu} \in k^*\langle\xi\rangle$ and $\eta \in k(\xi)$. Then $\eta \in k^*\langle\xi\rangle$ or $p = 2$, $\zeta_4 \in k^*\langle\xi\rangle$ and $\zeta_4 \notin k$.

Proof. By induction on $\nu$. For $\nu = 1$ the corollary follows from the statement $B_s$ above. Indeed, assume that $\xi^{p^{\mu-1}} \notin k^*$, $\eta \notin k^*\langle\xi\rangle$ and take in the proof of Theorem 25

$$M = k^*\langle\xi, \eta\rangle = N_{\mu+1}, \quad N_s = k^*\langle\xi^{p^{\mu-s}}\rangle \quad (0 \le s \le \mu).$$

We have $\eta^p \in N_\mu$, $\eta \in k(\xi)$. If $p > 2$, then by $B_\mu$ we get $\eta \in k^*\langle\xi\rangle$. Since the latter relation follows from $\eta \notin k^*\langle\xi\rangle$ it must be true.

If $p = 2$ and $1 + \zeta_4 \in M$ implies $\zeta_4 \in k$ then by $B_\mu$ again $\eta \in k^*\langle\xi\rangle$. Otherwise $1 + \zeta_4 \in k^*\langle\xi, \eta\rangle$ and $\zeta_4 \notin k$. Hence $2\zeta_4 = (1 + \zeta_4)^2 \in k^*\langle\xi, \eta^2\rangle$ and since $\eta^2 \in k^*\langle\xi\rangle$ we get $2\zeta_4 \in k^*\langle\xi\rangle$.

The proof for $\nu = 1$ is complete.

Suppose the corollary true for $\nu - 1$ $(\nu \ge 1)$ and let $\eta^{p^\nu} \in k^*\langle\xi\rangle$. From $(\eta^{p^{\nu-1}})^p \in k^*\langle\xi\rangle$, $\eta^{p^{\nu-1}} \in k(\xi)$, we get either $\eta^{p^{\nu-1}} \in k^*\langle\xi\rangle$ or $p = 2$ and $\zeta_4 \in k^*\langle\xi\rangle$, $\zeta_4 \notin k$. In the former case we apply the inductive assumption and thus get the corollary for all $\nu$.

We now proceed to prove an extension of Capelli's theorem.

Theorem 26. *Let $k$ be a field, $\alpha_i$ non-zero elements of $k$ $(i \le l)$, $n_i$ $(i \le l)$ positive integers at most one of them divisible by the characteristic of $k$, $\xi_i^{n_i} = \alpha_i$. Then*

$$(6) \qquad [k(\xi_1, \ldots, \xi_l) : k] = n_1 \ldots n_l$$

*if and only if*

$(7_1)$ *for all primes* $p$ *whenever*

$p|n_ix_i$ *for all* $i \le l$ *and* $\prod_{i=1}^{l} \alpha_i^{x_i} = \gamma^p$ *with* $\gamma \in k$ *then*

$$x_i \equiv 0 \pmod{p} \quad \textit{for all} \quad i$$

*and*

$(7_2)$ *whenever* $4|n_ix_i$ *for all* $i \le l$ *and*

$\prod_{i=1}^{l} \alpha_i^{x_i} = -4\gamma^4$ *with* $\gamma \in k$ *then* $x_i \equiv 0 \pmod{4}$

*for all* $i$.

In the case, where $n_1 \dots n_\ell \not\equiv 0 \bmod \operatorname{char} k$ the theorem was proved in Schinzel 1975b.

<u>Proof</u>. $(6) \longrightarrow (7_1)$ & $(7_2)$. Assume first that none of the numbers $n_i$ is divisible by char k, $\prod_{i=1}^{\ell} \alpha_i^{x_i} = \gamma^p$ gives $\prod_{i=1}^{\ell} \xi_i^{n_ix_i} = \gamma^p$ thus

$$\prod_{i=1}^{\ell} \xi_i^{n_ix_i/p} = \gamma\zeta_p^j .$$

If $j = 0$ and for some $i$ we have $p|n_i$, $p \nmid x_i$ then $\xi_i^{n_i/p} \in k^*\langle\xi_1,\dots,\xi_{i-1},\dots,\xi_\ell\rangle$ and

$[k(\xi_1,\dots,\xi_\ell) : k] < n_1 \dots n_\ell$.

If $j \ne 0$ the $\zeta_p \in k^*\langle\xi_1,\dots,\xi_\ell\rangle$ and by Theorem 25 applied with $M = k^*<\xi_1,\dots,\xi_\ell>$ we have $\xi_p \in k$, thus the previous case applies.

Similarly $\prod_{i=1}^{\ell} \alpha_i^{x_i} = -4\gamma^4$ gives $\prod_{i=1}^{\ell} \xi_i^{n_i x_i} = -4\gamma^4$, hence

$$\prod_{i=1}^{\ell} \xi_i^{n_i x_i/4} = \zeta_4^j(1+\zeta_4)\gamma. \tag{8}$$

Since

$$\zeta_4^j(1+\zeta_4) = \begin{cases} \pm(1+\zeta_4) & \text{if } j \equiv 0 \bmod 2, \\ \pm 2(1-\zeta_4) & \text{if } j \equiv 1 \bmod 2 \end{cases}$$

we get $1+\zeta_4 \in k^*<\xi_1,\dots,\xi_\ell>$ and by Theorem 25 $\zeta_4 \in k$. If for some $i$ we have $2|n_i$, $x_i \not\equiv 0 \bmod 4$ then by (8)

$\xi_i^{n/2} \in k^*<\xi_1,\dots,\xi_{i-1},\xi_{i+1},\dots,\xi_\ell>$ and

$[k(\xi_1,\dots,\xi_\ell) : k] < n_1\dots n_\ell$.

Now assume that one the numbers $n_i$, say $n_1$, is divisible by the characteristic $q$ of $k$ and let $n_1 = q^\nu n_1'$, where $n_1' \not\equiv 0 \bmod q$.
Clearly $[k(\xi_1^{q^\nu},\xi_2,\dots,\xi_\ell) : k] = n_1'n_2\dots n_\ell$ and applying the already proved assertion to $\alpha_1^{q^\nu},\alpha_2,\dots,\alpha_\ell$ we get the condition $(7_1)$ for all primes $p$ except $p = q$ and the condition $(7_2)$, which is trivially satisfied for $q = 2$.

The condition $(7_1)$ for $p = q$ takes the form

(9) $\alpha_1^{x_1} = \gamma^q$ implies $x_1 \equiv 0 \bmod q$

and it is indeed satisfied. Otherwise we would have $\alpha_1 = \gamma_1^q$ with $\gamma_1 \in k$ whence

$$\xi_1^{n/q} \in k \quad \text{and} \quad [k(\xi_1,\ldots,\xi_\ell) : k] < n_1 \ldots n_\ell .$$

$(7_1)$ & $(7_2) \longrightarrow (6)$. Again assume first that none of the numbers $n_i$ is divisible by $q = \text{char } k$. We need the following lemma, the easy proof of which is left to the reader.

Lemma. If $a_i, b_i \in \mathbb{Z}$ $(i \leq \ell)$, $(a_i, b_i) = 1$ and $b_i | m$, then

$$(m\frac{a_1}{b_1},\ldots,m\frac{a_\ell}{b_\ell}) = m\frac{(a_1,\ldots,a_\ell)}{[b_1,\ldots,b_\ell]} .$$

We proceed to verify the conditions (i) and (ii) of Theorem 25 with $M = k^*\langle\xi_1,\ldots,\xi_\ell\rangle$. Suppose that $\zeta_p \in k^*M$, i.e.

$$(10) \qquad \zeta_p = \gamma \prod_{i=1}^{\ell} \xi_i^{x_i} \quad \text{with} \quad \gamma \in k$$

and let

$$(11) \qquad m = [\frac{n_1}{(n_1,x_1)},\ldots,\frac{n_\ell}{(n_\ell,x_\ell)}] .$$

If $p|m$, raising both sides of (10) to power $m$ we get

$$1 = \gamma^m \prod_{i=1}^{\ell} \xi_i^{mx_i} = \gamma^m \prod_{i=1}^{\ell} \alpha_i^{mx_i/n_i} .$$

Since $p|m$ we have by $(7_1)$

$$\frac{mx_i}{n_i} \equiv 0 \pmod p \quad \text{for all} \quad i \le \ell .$$

Therefore,

$$\underset{1\le i\le \ell}{\text{g.c.d.}} \ \frac{mx_i}{n_i} = \underset{1\le i\le \ell}{\text{g.c.d.}} \ \frac{mx_i/(x_i,n_i)}{n_i/(x_i,n_i)} \equiv 0 \pmod p$$

and by the lemma and (11)

$$\underset{1\le i\le \ell}{\text{g.c.d.}} \ \frac{x_i}{(x_i,n_i)} \equiv 0 \pmod p .$$

Hence for an $i \le \ell$ $p\left|\frac{x_i}{(x_i,n_i)}\right.$ and $p\left|\frac{n_i}{(x_i,n_i)}\right.$, a contradiction. Thus $p \nmid m$ and from

$$\zeta_p^m = \gamma^m \prod_{i=1}^{\ell} \alpha_i^{mx_i/n_i} \in k$$

we get $\zeta_p \in k$.

The condition (i) of Theorem 25 is fulfilled. In order to prove (ii) suppose $1 + \zeta_4 \in k^*M$, i.e.

$$(12) \qquad 1 + \zeta_4 = \gamma \prod_{i=1}^{\ell} \alpha_i^{x_i} \quad \text{with} \quad \gamma \in k$$

and let again $m$ be given by (11).

If $4|m$, raising both sides of (12) to power $m$ we get

$$(-4)^{m/4} = \gamma^m \prod_{i=1}^{\ell} \alpha_i^{mx_i/n_i}$$

If $m/4$ is even the left hand side is a square in $k$ hence by $(7_1)$ $mx_i/n_i \equiv 0 \pmod 2$ for all $i \leq \ell$ and contradiction follows from the lemma as before.

If $m/4$ is odd, $m/4 = 2k+1$ we have

$$(-4)^{m/4} = -4 \cdot 2^{4t}$$

thus by $(7_2)$ $mx_i/n_i \equiv 0 \pmod 4$ for all $i \leq \ell$, which is again impossible by the lemma. Thus $4 \nmid m$ and from

$$(1+\zeta_4)^m = \gamma^m \prod_{i=1}^{\ell} \alpha_i^{mx_i/n_i}$$

we infer $\zeta_4 \in k$ since $2^{-[m/2]}(1+\zeta_4)^m = \pm\zeta_4$ or $\pm 1 \pm \zeta_4$.

Therefore the conditions of Theorem 25 are fulfilled and we have

$$[k(\xi_1,\ldots,\xi_\ell) : k] = [k^*\langle\xi_1,\ldots,\xi_\ell\rangle : k^*].$$

If the index on the right hand side were less than $n_1 \ldots n_\ell$, we would have

$$\prod_{i=1}^{\ell} \xi_i^{x_i} = \gamma \in k, \tag{13}$$

where for an $i \leq \ell$

$$x_i \not\equiv 0 \bmod n. \tag{14}$$

Raising both sides of (13) to power $m$, where $m$ is given by (11) we get

$$(15) \qquad \prod_{i=1}^{\ell} \alpha_i^{mx_i/n_i} = \gamma^m .$$

Take any prime divisor $p$ of $n_i/(n_i,x_i)$ where $i$ satisfies (14).

Clearly $p|m$. Applying to (15) the condition $(7_1)$ we get

$\underset{1\le i\le \ell}{\text{g.c.d.}} \dfrac{mx_i}{n_i} \equiv 0 \pmod p$. The lemma gives

$\underset{1\le i\le \ell}{\text{g.c.d.}} \dfrac{x_i}{(x_i,n_i)} \equiv 0 \pmod p$, a contradiction.

Now assume that char $k = q|n_1$, $n_1 = q^\nu n_1', n_1' \not\equiv 0 \bmod q$. The already proved assertion applied to $\xi_1^{q^\nu},\xi_2,\dots,\xi_\ell$ gives

$$(16) \qquad [k(\xi_1^{q^\nu},\xi_2,\dots,\xi_\ell) : k] = n_1'n_2,\dots,n_\ell .$$

On the other hand, since by (9) $\alpha_1^{n_1'} \ne \gamma^q$ we have by Theorem 21 $[k(\xi_1^{n_1'}) : k] = q^\nu$, hence

$$[k(\xi_1,\xi_2,\dots,\xi_\ell) : k] \equiv 0 \bmod q^\nu .$$

Combining this with (16) we get $[k(\xi_1,\dots,\xi_\ell) : k] = n_1n_2\dots n_\ell$ and the proof is complete.

Remark 1. Theorem 26 really includes Theorem 21 since if $\ell = 1$ the condition $(7_1)$ resolves into

$$(17) \qquad \alpha_1 \ne \gamma^p \quad \text{for all primes } p|n_1$$

and the condition (7) takes the form

$$\alpha_1^2 \neq -4\gamma^4 \quad \text{if} \quad 2|n_1 \quad \text{and besides} \quad \alpha_1 \neq -4\gamma^4 \text{ if } 4|n_1.$$

However the inequality $\alpha_1^2 \neq -4\gamma^4$ is trivially satisfied if $\zeta_4 \notin k$ and is a consequence of (17) if $\zeta_4 \in k$.

Remark 2. If at least two of the numbers $n_i$ are divisible by the characteristic of $k$ Theorem 25 may fail. An example is $k = \mathbb{F}_2(t)$, $\alpha_1 = t$, $\alpha_2 = t+1$, $n_1 = n_2 = 2$. Conditions (6) and (7) are satisfied, nevertheless $k(\xi_1,\xi_2) = k(\sqrt{t},\sqrt{t+1}) = k(\sqrt{t})$, thus

$[k(\xi_1,\xi_2) : k] = 2 < n_1 n_2$.

Remark 3. The following condition weaker than (7): $\zeta_4 \in k$ or $n_i x_i \equiv 0 \bmod 4$ for all $i \leq \ell$ implies $\prod_{i=1}^{\ell} \alpha_i^{x_i} \neq -4\gamma^4$ is necessary for the existence of $\xi_i$ satisfying

$$\xi_i^{x_i} = \alpha_i \quad \text{and} \quad [k(\xi_1,\ldots,\xi_\ell) : k] = [k^*\langle\xi_1,\ldots,\xi_\ell\rangle : k^*].$$

A sufficient condition is $\zeta_4 \in k$ or $n_i x_i \equiv 0 \bmod 4$ for all $i \leq \ell$ implies $\prod_{i=1}^{\ell} \alpha_i^{x_i} \neq -\gamma^4, -4\gamma^4$.

(see Schinzel 1975b). A simple necessary and sufficient condition is unknown to the writer.

# Part II. Arithmetic

## Section 18. A refinement of Gourin's theorem for algebraic number fields

<u>Definition 13.</u> *A cyclotomic polynomial over a field* $K$ *is a monic polynomial irreducible over* $K$ *whose zeros are roots of unity.*

<u>Definition 14.</u> *An extended cyclotomic polynomial over a field* $K$ *is a polynomial irreducible over* $K$ *of the form* $JF_0(x_1^{\delta_1} x_2^{\delta_2} \cdots x_s^{\delta_s})$, *where* $F_0$ *is a cyclotomic polynomial over* $K$ *and* $\delta_1, \delta_2, \dots, \delta_s$ *are integers not all equal to zero.*

If $K = \mathbf{Q}$, a cyclotomic polynomial over $K$ is simply called a cyclotomic polynomial and if it has $\zeta_m$ for a zero then the $m^{th}$ cyclotomic polynomial.

We can now formulate a refinement of Theorem 24.

<u>Theorem 27.</u> *Let* $K$ *be an algebraic number field and* $F(x_1,\dots,x_s)$ *a polynomial irreducible over* $K$, *not a constant multiple of any* $x_j$ $(j \le s)$ *or of any extended cyclotomic polynomial over* $K$. *Then for a suitable number* $c(F,K)$ *and for every vector* $\mathbf{t} = \langle t_1,\dots,t_s\rangle \in \mathbb{N}^s$ *there exist integral vectors* $\langle \tau_1,\dots,\tau_s\rangle \in \mathbb{N}^s$, $\langle u_1,\dots,u_s\rangle \in \mathbb{N}^s$ *such that*

$$\tau_i \le c(F,K),\ t_i = \tau_i u_i \quad (1 \le i \le s)$$

*and if* $F(x_1^{\tau_1},\dots,x_s^{\tau_s}) \overset{can}{\underset{K}{=}} \prod_{\rho=1}^{r} F_\rho(x_1,\dots,x_s)$

*then* $F(x_1^{t_1},\dots,x_s^{t_s}) \overset{can}{\underset{K}{=}} \prod_{\rho=1}^{r} F_\rho(x_1^{u_1},\dots,x_s^{u_s})$

Convention 1. Let $\alpha \in K^*$. If $\alpha^n = 1$ for some $n > 0$ then $e(\alpha,K) = 0$, if $\alpha^n \neq 1$ for all $n > 0$ then

$$e(\alpha,K) = \sup\{e : \alpha\beta^{-e} = \text{root of unity for some } \beta \in K\}.$$

Lemma 1. For all $\alpha \in K^*$, $e(\alpha,K) < \infty$.

Proof. If $\alpha$ is not a unit at least one prime ideal divides $\alpha$ with an exponent $a \neq 0$. If $\alpha = \zeta_q\beta^e$ we get $e|a$, whence $e < \infty$.

If $\alpha$ is a unit not a root of unity it has a factorization into fundamental units $\alpha = \zeta_w^{a_0}\varepsilon_1^{a_1}\ldots\varepsilon_n^{a_n}$ in which $n > 0$ and some $a_i \neq 0$. Then $\alpha = \zeta_q\beta^e$ implies $\beta$ is a unit and $e|(a_1,\ldots,a_n)$ whence $e < \infty$.

Lemma 2. For any positive integer $m$ and any $\alpha \in K^*$ we have $e(\alpha^m,K) = me(\alpha,K)$.

Proof. If $\alpha$ is a root of unity we have $e(\alpha^m,K) = 0 = me(\alpha,K)$. Therefore, assume $\alpha$ is not a root of unity.

Suppose $\alpha = \zeta_w^{j_1}\beta_1^{e_1}$, $\alpha = \zeta_w^{j_2}\beta_2^{e_2}$. If $(e_1,e_2) = d$, $[e_1,e_2] = e$ we have $d = ae_1 + be_2$,

$$\alpha = (\zeta_w^{j_1}\beta_1^{e_1})^{be_2/d}(\zeta_w^{j_2}\beta_2^{e_2})^{ae_1/d} = \zeta_w^j(\beta_1^b\beta_2^a)^e.$$

It follows that $e(\alpha,K)$ is the least common multiple of all integers $e$ satisfying $\alpha = \zeta_q\beta^e$ for some $\beta \in K$ and $q \in \mathbb{N}$.

Applying this observation to $\alpha^m$ we get $me(\alpha,K) | e(\alpha^m,K) = \tilde{e}$.

On the other hand

$$\alpha^m = \zeta_w^j \beta^{\tilde{e}} \quad \text{implies} \quad \alpha = \zeta_q \beta^{\tilde{e}/m} \quad \text{thus} \quad e(\alpha,K) \equiv 0 \bmod \frac{\tilde{e}}{m}.$$

Since $\tilde{e}$ and $me(\alpha,K)$ divide each other we get the lemma.

<u>Convention 2</u>. Let $\alpha \in K^*$. If $\alpha^n = 1$ for some $n > 0$ then

$E(\alpha,K) = 0$, if $\alpha^n \neq 1$, for all $n > 0$, then

$E(\alpha,K) = \sup\{E : \alpha = \theta^E$ for some $\theta \in K(\zeta_E)\}$.

<u>Lemma 3</u>. For all $\alpha \in K^*$ we have $E(\alpha,K) < \infty$.

<u>Proof</u>. Assume without loss of generality that $\zeta_4 \in K$ and

$\alpha$ is not a root of unity. Let

$$\alpha = \theta^E, \quad E = \prod_{i=1}^{n} p_i^{a_i}, p_1 < p_2 < \dots < p_n, \quad E = p_i^{a_i} E_i = q_i E_i.$$

In Corollary 4 to Theorem 25 take

$$p = p_i, \quad k = K(\zeta_{E_i}), \quad \mu = \nu = a_i,$$

$$\xi = \zeta_{q_i}, \eta = \theta^{E_i}.$$

Since $\xi^{p^\mu} = \zeta_{q_i}^{q_i} = 1 \in k$,

$$\eta = \theta^{E_i} \in K(\zeta_E) = k(\xi),$$

$$\eta^{p^\nu} = \theta^{E_i q_i} = \alpha \in k$$

we have by that corollary that

$$\theta^{E_i} \in k^*\langle\xi\rangle .$$

Hence $\theta^{E_i} = \beta\zeta_{q_i}^j$, $\beta \in k^*$,

$$(1) \qquad \alpha = \theta^{E_i q_i} = \beta^{q_i}.$$

Taking norms $N$ from $k$ to $K$ we get

$$N = \alpha^{[K(\zeta_{E_i}):K]} = N\beta^{q_i}.$$

Since

$$[K(\zeta_{E_i}):K] \mid [\emptyset(\zeta_{E_i}):\emptyset] \quad \text{and} \quad [\emptyset(\zeta_{E_i}):\emptyset] = \phi(E_i)$$

we get

$$\alpha^{\phi(E_i)} = \gamma^{q_i}, \qquad \gamma \in K.$$

Lemma 2 gives $e(\alpha^{\phi(E_i)},K) = \phi(E_i)e(\alpha,K) \equiv 0 \bmod q_i$ by (1), hence

$$(2) \quad p_i^{a_i} \mid \phi(E_i)e(\alpha,K), \qquad (1 \le i \le n).$$

Since $(p_n^{a_n},\phi(E_n)) = 1$, when $i = n$ we have

$p_n^{a_n} | e(\alpha,K)$, $p_n \le e(\alpha,K)$.

Further (2) implies

$$p_i^{a_i} | (p_1 - 1)(p_2 - 1)\dots(p_n - 1)e(\alpha,K).$$

Hence in view of

$$p_1 < p_2 < \dots < p_n \le e(\alpha,K)$$

we have

$$p_i^{a_i} | e(\alpha,K)! \quad \text{and} \quad E | e(\alpha,K)!.$$

Thus $E(\alpha,K) | e(\alpha,K)!$ and the conclusion follows from Lemma 1.

Remark. For a more precise estimate of $E(\alpha,K)$ in terms of $e(\alpha,K)$ see Schinzel 1978.

Lemma 4. If $\alpha^m \in K$, $\alpha \in K(\zeta_m)$, then

$$E(\alpha^m,K) \equiv 0 \bmod mE(\alpha,K(\zeta_m)).$$

Proof. Suppose $\alpha = \theta_1^{E_1}$, $\theta_1 \in K(\zeta_{E_1})$,

$$\alpha = \theta_2^{E_2}, \quad \theta \in K(\zeta_{E_2}),$$

$D = (E_1, E_2)$, $E = [E_1, E_2]$. We have

$$D = aE_1 + bE_2 ,$$

$$\alpha = (\theta_1^{E_1})^{bE_2/D} (\theta_2^{E_2})^{aE_1/D}$$

$$= (\theta_1^b \theta_2^a)^E , \quad \theta_1^b \theta_2^a \in K(\zeta_E).$$

It follows that $E(\alpha,K)$ is the least common multiple of all exponents $E$ with the property that $\alpha = \theta^E$, $\theta \in K(\zeta_E)$.

Now let $\alpha = \theta^E$, $\theta \in K(\zeta_m, \zeta_E)$. Then

$$\alpha^m = \theta^{Em} \text{ with } \theta \in K(\zeta_m).$$

The lemma follows on applying the previous observation to $\alpha^m$.

Lemma 5. Let $n \in \mathbb{N}$, $F \in K[x]$, $\alpha \in K^*$, $\nu = (n, E(\alpha,K))$. Then if $F(x) \mid x^n - \alpha$ in $K[x]$ and $F(x)$ is irreducible over $K$, then

$$F(x) = F_o(x^{n/\nu}) \text{ with } F_o(x) \mid x^\nu - \alpha, \quad F_o \in K[x].$$

Proof. By induction with respect to $E(\alpha,K)$.

If $E(\alpha,K) = 0$ then $\nu = n$, the lemma holds trivially with $F_o = F$.

Suppose the lemma proved for all $\alpha$ with $E(\alpha,K) < N$ and let $e(\alpha,K) = N$.

If $x^n - \alpha$ is irreducible then $F(x) = \text{const}(x^n - \alpha)$ and the lemma holds with $F_o = \text{const}(x^\nu - \alpha)$.

Suppose $x^n - \alpha$ reducible. Theorem 21 gives either

$$\alpha = \beta^p, \quad \beta \in K \quad \text{and} \quad p|n,$$

or $$\alpha = -4\beta^4, \quad \beta \in K \quad \text{and} \quad 4|n.$$

In the first case

$$F(x)|x^n - \alpha = (x^{n/p} - \beta)\prod_{j=1}^{p-1}(x^{n/p} - \zeta_p^j\beta).$$

There are two possibilities:

1) $F(x)|x^{n/p} - \beta$. By Lemma 4

$E(\beta,K)|\frac{1}{p}E(\alpha,K)$. Hence by the inductive assumption

$$F(x) = G_o(x^{n/p\nu_o}), \quad \nu_o = (\frac{n}{p}, E(\beta,K)), \quad G_o(x)|x^{\nu_o} - \beta.$$

We have $\nu_o|\frac{\nu}{p}$ and $G_o(x^{\nu/p\nu_o})|x^{\nu/p} - \beta|x^\nu - \alpha$. It suffices to take $F_o = G_o(x^{\nu/p\nu_o})$.

2) $F(x)|\prod_{j=1}^{p-1}(x^{n/p} - \zeta_p^j\beta)$.

Let $\Phi(x)$ be a factor of $F$ irreducible over $K(\zeta_p)$ and monic. We have

$$\Phi(x)|x^{n/p} - \zeta_p^j\beta \quad \text{for some} \quad j < p.$$

Conjugate factors divide conjugate expressions, hence the conjugates of $\Phi$ over $K(x)$ are pairwise relatively prime. Hence $F = \text{const}\ N_{K(\zeta_p)/K}\Phi$.

By Lemma 4

$$E(\zeta_p^j\beta, K(\zeta_p)) \mid \frac{1}{p} E(\alpha, K)$$

thus by the inductive assumption

$$\Phi(x) = \Phi_o(x^{n/p\nu_o}), \quad \text{where} \quad \Phi_o \mid x^{\nu_o} - \zeta_p^j\beta, \quad \nu_o = (\tfrac{n}{p}, E(\zeta_p^j\beta, K(\zeta_p))).$$

Hence $F = \text{const}\ N_{K(\zeta_p)/K}\Phi_o(x^{n/p\nu_o})$ and it suffices to take $F_o = N_{K(\zeta_p)/K}\Phi_o(x^{\nu/p\nu_o})$.

In the second case $\alpha = -4\beta^4$, $4 \mid n$. We have

$$F(x) \mid x^n - \alpha = \prod_{j=0}^{3} (x^{\frac{n}{4}} - \zeta_4^j(1+\zeta_4)\beta).$$

Let $\Phi(x)$ be a factor of $F$ irreducible over $K(\zeta_4)$ and monic. For some $j < 4$ we have $\Phi(x) \mid x^{\frac{n}{4}} - \zeta_4^j(1+\zeta_4)\beta$.

Conjugate factors divide conjugate expressions, hence the possible conjugate of $\Phi$ over $K(x)$ is prime to $\Phi$ and $F = \text{const}\ N_{K(\zeta_4)/K}\Phi$. By Lemma 4 $E(\zeta_4^j(1+\zeta_4), K(\zeta_4)) \mid \frac{1}{4}E(\alpha, K)$ and by the inductive assumption

$$\Phi(x) = \Phi_o(x^{n/4\nu_o}), \quad \Phi_o(x) \mid x^{n/4} - \zeta_4^d(1+\zeta_4)\beta,$$

$$\nu_o = (\tfrac{n}{4}, E(\zeta_4^j(1+\zeta_4)), K(\zeta_4)).$$

We take

$$F_o = \text{const}\ N_{K(\zeta_4)/K}\Phi_o(x^{\nu/p\nu_o}).$$

Lemma 6. Theorem 27 holds for polynomials in one variable $(s = 1)$.

Proof. The condition of the theorem means in this case that $F \neq cx$ and $F$ is not a constant multiple of a cyclotomic polynomial. Let $\alpha$ be a zero of $F$. Then

$F(x) = \text{const}\ N_{K(\alpha)/K}(x - \alpha)$, thus $F(x^t) = \text{const}\ N_{K(\alpha)/K}(x^t - \alpha)$.

By Theorem 20, if

$$x^t - \alpha \overset{\text{can}}{\underset{K(\alpha)}{=}} \prod_{\rho=1}^{r} \Phi_\rho(x),$$

then
$$F(x^t) \overset{\text{can}}{\underset{K}{=}} \prod_{\rho=1}^{r} N_{K(\alpha)/K}\Phi_\rho(x).$$

Let us take

$$\tau = (t, E(\alpha, K(\alpha)).$$

We assert that

$$F(x^\tau) \overset{\text{can}}{\underset{K}{=}} \prod_{\rho=1}^{r} F_\rho(x) \tag{3}$$

implies

$$(4) \qquad F(x^t) \underset{K}{\overset{\text{can}}{=}} \prod_{\rho=1}^{r} F_\rho(x^{t/\tau}).$$

If (3) holds the factors on the right hand side of (4) are coprime by Lemma 1 to Theorem 24. Suppose

$$(5) \qquad G_\rho(x) \mid F_\rho(x^{t/\tau}), \quad G_\rho \quad \text{irreducible over} \quad K.$$

By (3) $G_\rho(x) \mid F(x^t)$ hence $\text{const}\ G_\rho(x) = N_{K(\alpha)/K}\Phi_\rho(x)$, where $\Phi_\rho(x)$ is irreducible over $K(\alpha)$ dividing $x^t - \alpha$. By the preceding lemma with $K$ replaced by $K(\alpha)$, $n$ by $t$

$$\Phi_\rho(x) = \Psi(x^{t/\tau}), \quad \Psi(x) \mid x^\tau - \alpha, \quad \Psi \in K(\alpha)[x].$$

Then $\text{const}\ G_\rho(x) = N_{K(\alpha)/K}\Psi(x^{t/\tau})$ and by (5) $N_{K(\alpha)/K}\Psi(x) \mid F_\rho(x)$. Since $F_\rho(x)$ is irreducible we have $N_{K(\alpha)/K}\Psi(x) = \text{const}\ F_\rho(x)$, thus $G_\rho(x) = \text{const}\ F_\rho(x^{t/\tau})$. Since $G_\rho$ is irreducible over $K$ the lemma follows.

<u>Lemma 7</u>. If $(b_1, \ldots, b_s) = 1$ and $F_o(x)$ is irreducible over $K$, $F_o(x) \neq cx$ then $JF_o(x_1^{b_1} \ldots x_s^{b_s})$ is irreducible over $K$.

<u>Proof</u>. Let $F_o(\alpha) = 0$, $F_o = \text{const}\ N_{K(\alpha)/K}(x - \alpha)$. Then

$$JF_o(x_1^{b_1} \ldots x_s^{b_s}) = \text{const}\ N_{K(\alpha)/K}J(x_1^{b_1} \ldots x_s^{b_s} - \alpha).$$

By Theorem 21 $\Phi = J(x_1^{b_1} \ldots x_s^{b_s} - \alpha)$ is irreducible over $\hat{K}$.

By Lemma 2 to Theorem 24, $N_{K(\alpha)/K}\Phi$ is irreducible over K. This proves the lemma.

<u>Corollary</u>. F is an extended cyclotomic polynomial if and only if $F = JF_0(x_1^{\delta_1} \dots x_s^{\delta_s})$, where $F_0$ is a cyclotomic polynomial over K and $(\delta_1,\dots,\delta_s) = 1$.

<u>Proof</u>. The lemma implies that the condition is sufficient. In order to prove the necessity, suppose that $F = JF_0(x_1^{\delta_1} \dots x_s^{\delta_s})$, where $F_0$ is a cyclotomic polynomial over K and $(\delta_1,\dots,\delta_s) = \delta$. Since F is irreducible in K, $F_1(x) = F_0(x^{\delta})$ is also. On the other hand if $F_0(\zeta_m) = 0$ we have $F_1(\zeta_{m\delta}) = 0$ thus $F_1$ is a cyclotomic polynomial and $F = JF_1(x_1^{\delta_1/\delta} \dots x_s^{\delta_s/\delta})$, where now $(\delta_1/\delta,\dots,\delta_s/\delta) = 1$.

<u>Proof of Theorem 27</u>. If $F \neq JF_0(x_1^{\delta_1} \cdot \dots \cdot x_s^{\delta_s})$ for any $F_0$ then Theorem 24 implies Theorem 27.

If $F = JF_0(x_1^{\delta_1} \cdot \dots \cdot x_s^{\delta_s})$, $F_0$ non-cyclotomic we take $\alpha$ a zero of $F_0$,

$$c(F,K) = E\,(\alpha,K(\alpha)),$$

$$\tau_j = (t_j,c(F,K)), \quad u_j = t_j\tau_j^{-1}.$$

If $\delta(\mathbf{t}) = (t_1\delta_1,\ldots,t_s\delta_s)$ we have $(\frac{t_1\delta_1}{\delta(\mathbf{t})},\ldots,\frac{t_s\delta_s}{\delta(\mathbf{t})}) = 1$.

By Lemma 6 if $\delta = (\delta(\mathbf{t}),c(F,K))$ then

$$F_0(x^\delta) \underset{K}{\overset{\text{can}}{=}} \prod_{\rho=1}^{r} G_\rho(x) \quad \text{implies} \quad F_0(x^{\delta(\mathbf{t})}) \underset{K}{\overset{\text{can}}{=}} \prod_{\rho=1}^{r} G_\rho(x^{\delta(\mathbf{t})/\delta}).$$

Hence by Lemma 7 the polynomials $JG_\rho(x_1^{\delta_1 t_1/\delta},\ldots,x_s^{\delta_s t_s/\delta})$ $(\rho \leq r)$ are irreducible. Moreover by Lemma 3 to Theorem 24 they are pairwise relatively prime. Since $\delta_j\tau_j/\delta =$ $=(\delta_j t_j,\delta_j c(F,K))/\delta \in \mathbf{Z}$ and $\delta_j\tau_j | \delta_j t_j$ the same applies to the polynomials $F_\rho = JG_\rho(x_1^{\delta_1 t_1/\delta},\ldots,x_s^{\delta_s t_s/\delta})$.

Thus $F(x_1^{\tau_1},\ldots,x_s^{\tau_s}) \underset{K}{\overset{\text{can}}{=}} \prod_{\rho=1}^{r} F_\rho(x_1,\ldots,x_s)$,

$F(x_1^{t_1},\ldots,x_s^{t_s}) \underset{K}{\overset{\text{can}}{=}} \prod_{\rho=1}^{r} F_\rho(x_1^{u_1},\ldots,x_s^{u_s})$ and the theorem follows.

*Corollary*. Let $\Omega(G,K)$ be the number of irreducible factors of a polynomial $G$ over $K$ counting multiplicity. Then under the assumptions of Theorem 27 there exist a constant $c_1(F,K)$ such that

$$\Omega(F(x_1^{t_1},\ldots,x_s^{t_s}),K) < c_1(F,K).$$

On the other hand if $\Phi_m(x)$ is the m-th cyclotomic polynomial, then for $n$ prime to $m$ we have

$$\Phi_m(x^n) = \prod_{d|n} \Phi_{md}(x), \quad \text{thus}$$

$$J\Phi_m(x_1^{\delta_1 n} \ldots x_s^{\delta_s n}) = \prod_{d|n} J\Phi_{md}(x_1^{\delta_1} \ldots x_s^{\delta_s})$$

has for suitable $n$ arbitrarily many factors irreducible over $\mathbb{Q}$.

We recall that for a given polynomial

$$F = \sum_{i_1,\ldots,i_s=0}^{|F|} a_{i_1\ldots i_s} x_1^{i_1} \ldots x_s^{i_s} \in \mathbb{C}[x_1,\ldots,x_s]$$

we set

$$\| F \| = \sum_{i_1,\ldots,i_s=0}^{|F|} |a_{i_1\ldots i_s}|^2.$$

Conjecture. There exist a function $\psi$ such that for every polynomial $F \in \mathbb{Z}[x_1,\ldots,x_s]$ prime to $x_1 \cdot \ldots \cdot x_s$ the number $\Omega_1(F)$ of factors of $F$ irreducible over $\mathbb{Q}$ that are not extended cyclotomic, counted with multiplicities, does not exceed $\psi(\| F \|)$.

The best known approach to this conjecture is via Smyth's theorem to be discussed in the next section.

## Section 19. Smyth's theorem on non-reciprocal polynomials

Definition 15. *A polynomial* $F(x_1,\ldots,x_k)$ *is reciprocal if*

$$JF(x_1^{-1},\ldots,x_s^{-1}) = \pm F(x_1,\ldots,x_s).$$

Definition 16. *For a given polynomial* $P = p_o \prod_{j=1}^{n} (z-\alpha_j) \in \mathbb{C}[z]$ *we set*

$$M(P) = |p_o| \prod_{j=1}^{n} \max(1,|\alpha_j|).$$

Theorem 28 (Smyth, 1971). *Let* $P \in \mathbf{Z}[z]$ *be a non-reciprocal polynomial with* $P(0) \neq 0$. *Then*

$$M(P) \geq \theta_o,$$

*where* $\theta_o$ *is the real zero of* $x^3 - x - 1$ $(\theta_o = 1.3247\ldots)$.

Lemma 1. Let $F(z) = \sum_{i=0}^{\infty} e_i z^i$ be a function holomorphic in an open disc containing $|z| \leq 1$ and such that $|f(z)| \leq 1$ on $|z| = 1$.

Then $|e_i| \leq 1 - |e_o|^2$. Moreover if $e_i$ are real, then

$$e_{2i} \leq 1 - e_o^2 - \frac{e_i^2}{1-e_o}$$

$$e_{2i} \geq -(1 - e_o^2 + \frac{e_i^2}{1+e_o}).$$

Proof. Let $f(z)(\beta + z^i) = e_o\beta + \ldots + (e_i\beta + e_o)z^i + \ldots$ . Applying Parseval's identity, and the inequality

$$|f(z)(\beta + z^i)| \leq |\beta + z^i| \quad \text{for} \quad |z| = 1, \quad \text{we get}$$

$$|e_o\beta|^2 + |e_i\beta + e_o|^2 \leq \int_{|z|=1} |\beta + z^i|^2 dz = |\beta|^2 + 1.$$

Choose $\beta = \dfrac{\bar{e}_i}{|e_i|\bar{e}_o}$. It follows that

$$\left|\frac{e_i\bar{e}_i}{|e_i|\bar{e}_o} + e_o\right| \leq \frac{1}{|e_o|} \; ; \quad |e_i| + |e_o|^2 \leq 1 \quad \text{and} \quad |e_i| \leq 1 - |e_o|^2.$$

The prove the second part we consider instead

$$f(z)(\beta_o + \beta_1 z^i + z^{2i}) = e_o\beta_o + \ldots + (e_o\beta_1 + e_i\beta_o)z^i + \ldots$$

$$+ \ldots (e_o + e_i\beta_1 + e_{2i}\beta_o)z^{2i} + \ldots \; .$$

Arguing as before, we get

$$|e_o\beta_o|^2 + |e_i\beta_o + e_o\beta_1|^2 + |e_o + e_i\beta_1 + e_{2i}\beta_o|^2$$

$$\leq \int_{|z|=1} |\beta_o + \beta_1 z^i + z^{2i}|^2 \, dz = \beta_o^2 + \beta_1^2 + 1.$$

Choose $\beta_o = \dfrac{\rho}{e_o}$, $\rho = \pm 1$, and $e_i\beta_o + e_o\beta_1 = \delta\beta_1$, $\delta = \pm 1$,

$$\beta_1 = \frac{e_i\beta}{\delta - e_o}.$$

The resulting inequality is

$$(e_o + \frac{e_i^2\beta_o}{\delta - e_o} + e_{2i}\beta_o)^2 \leq \beta_o^2 .$$

Hence

$$|e_o + \frac{e_i^2\rho}{(\delta - e_o)e_o} + \frac{e_{2i}\rho}{e_o}| \leq \frac{1}{|e_o|} ,$$

$$|\rho e_o + \frac{e_i^2}{(\delta - e_o)e_o} + \frac{e_{2i}}{e_o}| \leq \frac{1}{|e_o|} ,$$

$$|\rho e_o^2 + \frac{e_i^2}{\delta - e_o} + e_{2i}| \leq 1.$$

Since this holds for both values of $\rho$ we get

$$-1 + e_o^2 \leq e_{2i} + \frac{e_i^2}{\delta - e_o} \leq 1 - e_o^2 .$$

On the right we take $\delta = 1$, on the left $\delta = -1$ and the lemma follows.

Remark. If $e_i$ are complex one can show that

$$|e_{2i} + \frac{e_i^2\bar{e}_o}{1-|e_o|^2}| \leq 1 - |e_o|^2 - \frac{|e_i|^2}{1-|e_o|^2}$$

(see Bazylewicz 1976).

Lemma 2. Let $P(z) \in \mathbb{C}[z]$ have the leading coefficient $p_o$ and let $|p_o| = |P(0)|$. Then if $Q(z) = z^{|P|}\bar{P}(z^{-1})$ we have

$$\frac{\bar{P}(0)}{p_o} \frac{P(z)}{Q(z)} = \frac{f(z)}{g(z)} ,$$

where $f$ and $g$ are holomorphic in an open disc containing

$|z| \leq 1$ and

$$|f(0)| = |g(0)| = |p_o|M(P)^{-1}, \quad |f(z)| = |g(z)| = 1 \text{ on } |z| = 1.$$

Moreover if the coefficients of $P$ are real, then the Taylor coefficients of $f$ and $g$ are also real and $f(0) = g(0) > 0$.

Proof: Let $P(z) = p_o \prod_{j=1}^{n} (z - \alpha_j)$ and put

$$f(z) = \prod_{|\alpha_j|<1} \frac{z-\alpha_j}{1-\bar{\alpha}_j z} \cdot \prod_{|\alpha_j|=1} (-\alpha_j),$$

$$g(z) = \prod_{|\alpha_j|>1} \frac{1-\bar{\alpha}_j z}{z-\alpha_j}.$$

We have

$$\frac{f(z)}{g(z)} = \frac{\prod_{|\alpha_j|\neq 1} (z-\alpha_j)}{\prod_{|\alpha_j|\neq 1} (1-\bar{\alpha}_j z)} \cdot \prod_{|\alpha_j|=1} (-\alpha_j).$$

If $|\alpha_j| = 1$ then $\bar{\alpha}_j = \alpha_j^{-1}$, thus $\dfrac{z-\alpha_j}{1-\bar{\alpha}_j z} = \dfrac{z-\alpha_j}{\alpha_j - z}\alpha_j = (-\alpha_j)$.

Hence

$$\frac{f(z)}{g(z)} = \frac{\prod(z-\alpha_j)}{\prod(1-\bar{\alpha}_j z)} = \frac{\bar{P}(0)P(z)}{p_o Q(z)}.$$

Furthermore

$$|f(0)| = \prod_{|\alpha_j|<1} |\alpha_j| = |p_o| M(P)^{-1} = |g(0)|.$$

It remains to check $\overline{f(z)} = f(\bar{z})$ if $P,Q \in \mathbb{R}(z)$. Now in the latter case $\bar{\alpha}_j$ is conjugate to $\alpha_j$, hence

$$\overline{f(z)} = \overline{\prod_{|\alpha_j|=1} (-\alpha_j)} \prod_{|\alpha_j|<1} \frac{\bar{z} - \bar{\alpha}_j}{1-\alpha_j \bar{z}} = f(\bar{z}).$$

Similarly $\overline{g(z)} = g(\bar{z})$. We can make $f(0) = g(0) > 0$ by changing sign of $f$ and $g$ if necessary.

Proof of Theorem 28. If $|p_o P(0)| \neq 1$, then $M(P) \geq \max(|p_o|, |P(0)| \geq 2 > \theta_o$. Thus we can assume $p_o = \pm 1$, $P(0) = \pm 1$. Applying Lemma 2 we get

$$\frac{P(0)P(z)}{Q(z)} = \frac{f(z)}{g(z)} = 1 + a_k z^k + a_\ell z^\ell + \ldots, \quad a_k a_\ell \neq 0 (k < \ell)$$

$$a_k, a_\ell \in \mathbb{Z},$$

where the expansion on the right is infinite since $P(z)$ is not reciprocal.

Let $f(z) = c + c_1 z + c_2 z^2 + \ldots,$

$g(z) = d + d_1 z + d_2 z^2 + \ldots .$

We have $c = d = M(P)^{-1}$, $c_i = d_i \quad (i < k)$,

$$(1) \qquad \begin{cases} c_k = d_k + a_k c, \\ c_i = d_i + a_k c_{i-k} \qquad (k < i < \ell), \\ c_\ell = d_\ell + a_k c_{\ell-k} + a_\ell c. \end{cases}$$

By Lemma 1 it follows that

$$|c_k| \leq 1 - c^2 ,$$

$$|d_k| \leq 1 - c^2 ,$$

$$c_k - d_k = a_k c .$$

If $|a_k| \geq 2$ then $|c_k - d_k| \geq 2c$ hence $\max(|c_k|,|d_k|) \geq c$,

$c \leq 1 - c^2$, $c \leq \frac{-1+\sqrt{5}}{2}$ thus $c^{-1} \geq \frac{1+\sqrt{5}}{2} > \theta_o$.

It remains to consider $|a_k| = 1$, and $|c_k| + |d_k| = c$.

It is enough to consider the case $a_k = 1$. If $a_k = -1$ we replace $P(z)$ by $Q(z) = z^{|P|}P(z^{-1})$ and we have this case again since $M(Q) = M(P)$ and $\frac{1}{1-z^k+a\ z+\ldots} = 1+z^k+\ldots$ .

We can also assume $|a_\ell| = 1$ since if $|a_\ell| \geq 2$, then

$\max(|c_\ell|,|c_{\ell-k}|,|d_\ell|) \geq \frac{2c}{3}$ and we would get $\frac{2}{3}c \leq 1 - c^2$,

hence $c^{-1} > \theta_o$.

<u>Case 1</u>. $\ell < 2k$. Let us consider the functions

$$f(z)(1+\gamma z^{\ell-k}-z^k+\beta z^\ell)=c+(c\gamma+c_{\ell-k})z^{\ell-k}+(-c+\gamma c_{2k-\ell}+c_k)z^k$$

$$+(\beta c-c_{\ell-k}+\gamma c_k+c_\ell)z^\ell+\ldots$$

$$g(z)(-1-\gamma z^{\ell-k}-z^k+\beta z^\ell)=-c+(-c\gamma-d_{\ell-k})z^{\ell-k}+(-c-\gamma d_{2k-\ell}-d_\ell)z^k$$

$$+(\beta c-d_{\ell-k}-\gamma d_k-d_k)z^\ell+\ldots .$$

Applying to them Parseval's identity and using $|f(z)| \leq 1$, $|g(z)| \leq 1$ we get

$$c^2+(c\gamma+c_{\ell-k})^2+(-c+\gamma c_{2k-\ell}+c_k)^2+(\beta c-c_{\ell-k}+\gamma c_k+c_\ell)^2 \leq 2+\beta^2+\gamma^2,$$

$$c^2+(-c\gamma - d_{\ell-k})^2+(-c-\gamma d_{2k-\ell}-d_k)^2+(\beta c-d_{\ell-k}-\gamma d_k-d_\ell) \leq 2+\beta^2+\gamma^2.$$

Now $\frac{a^2+b^2}{2} \geq (\frac{a+b}{2})^2$, thus

$$c^2+(c\gamma+c_{\ell-k})^2+(-c+\frac{c_k-d_k}{2})^2+(\beta c-c_{\ell-k}+\gamma\frac{c_k-d_k}{2}+\frac{c_\ell-d_\ell}{2})^2\leq 2+\beta^2+\gamma^2.$$

Replacing, by virtue of (1) $c_k-d_k$ by $c$, $c_\ell-d_\ell$ by $c_{\ell-k}+a_\ell c$, we get:

$$(2)\quad \max_{\beta,\gamma}\{c^2+(c_{\ell-k}+\gamma c)^2+\frac{c^2}{4}+(\frac{c_{\ell-k}+a_\ell c}{2}+\frac{\gamma}{2}c-c_{\ell-k}+\beta c)^2-\beta^2-\gamma^2\}\leq 2.$$

Replacing in the expression in brackets $\beta,\gamma,c_{\ell-k}$ by $\frac{\beta}{t}, \frac{\gamma}{t}, \frac{c_{\ell-k}}{t}$ and multiplying through by $t^2$ we get a quadratic form $F(\beta,\gamma,c_{\ell-k},t)$. The matrix of this form is

$$\begin{pmatrix} c^2-1, & \frac{c^2}{2}, & -\frac{c}{2}, & \frac{a_\ell c^2}{2} \\ 0, & \frac{5}{4}c^2-1, & \frac{c}{4}, & \frac{a_\ell c^2}{2} \\ 0, & 0, & \frac{5}{4}c^2, & \frac{-a_\ell c}{4} \\ 0, & 0, & 0, & \frac{3}{2}c^2 \end{pmatrix}$$

Let us compute the leading minors:

$$M_1 = c^2 - 1,$$

$$M_2 = c^4 - \frac{9}{4} c^2 + 1,$$

$$M_3 = \frac{5}{4} - 2c^2,$$

$$M_4 = \frac{25}{16} c^2 - \frac{5}{2} c^4 + \frac{c^2 a_\ell^2}{4} .$$

As we shall presently show $M_1 M_2 M_3 \neq 0$, hence (cf.Dickson 1926, p. 27)

$$(3) \qquad F(\beta,\gamma,c_{\ell-k},1) = M_1(\beta+\dots)^2 + \frac{M_2}{M_1}(\gamma+\dots)^2 + \frac{M_3}{M_2}(c_{\ell-k}+\dots)^2 + \frac{M_4}{M_3} .$$

Now, $c = \prod_{|\alpha_i|>1} |\alpha_i|^{-1} < 1$, so $M_1 < 0$. Further

$|c_k| \leq 1 - c^2$, $|d_k| \leq 1 - c^2$, also $|c_k| + |d_k| = c$, thus

$$c \leq 2 - 2c^2,\ 2c^2 - c - 2 \leq 0,\quad c \leq \frac{-1+\sqrt{17}}{4} .$$

Since $c^{-1}$ is an algebraic integer, the equality is excluded, and we get

$$(4) \qquad c < \frac{-1+\sqrt{17}}{4}, \quad c^2 < \frac{9-\sqrt{17}}{8} .$$

Now $c^4 - \frac{9}{4} c^2 + 1 = (c^2 - \frac{9-\sqrt{17}}{8})(c^2 - \frac{9+\sqrt{17}}{8})$, hence $M_2 > 0$.

Also $M_3 > 0$, for otherwise $2c^2 > \frac{5}{4}$, $c^2 > \frac{5}{8}$ contrary to (4). Finally $M_4 \geq \frac{29}{16} c^2 - \frac{5}{2} c^4$.

By choosing $\gamma$ as a function of $c_{\ell-k}$ we can make the second term on the right hand of (3) equal to 0, choosing $\beta$ we can make the first term 0. Hence

$$\max_{\beta,\gamma} F(\beta,\gamma,c_{\ell-k},1) \geq \frac{M_4}{M_3},$$

and by (2) $\frac{M_4}{M_3} \leq 2$, or after reduction

$$40c^4 - 93c^2 + 40 \geq 0, \quad 40c^{-4} - 93c^{-2} + 40 \geq 0.$$

Therefore, it is enough to show

$$E = 40\,\theta_o^4 - 93\,\theta_o^2 + 40 < 0.$$

Since $\theta_o^4 = \theta_o^2 + \theta_o$, we have

$$E = 40\,\theta_o^4 - 53\,\theta_o^2 + 40 = 40\,\theta_o^3 - 53\,\theta_o^2 =$$

$$= 40\,\theta_o^2(\theta_o - \tfrac{53}{40}) < 40\,\theta^2(1.3248 - 1.3250) < 0.$$

In fact, in case 1, $M(P) = c^{-1} > \theta_o + .001$

Case 2. $\ell \geq 2k$. If $\ell = 2k$, we have

$$\frac{P(0)P(z)}{Q(z)} = 1 \pm z^k \pm z^{2k} + \ldots$$

Using the same trick as before, (replacing $P(z)$ by $Q(z)$) we get

$$\frac{Q(0)Q(z)}{P(z)} = \frac{1}{1\pm z^k \pm z^{2k} + \dots} = 1 \mp z^k + \begin{cases} 0 \cdot z^{2k} \\ \\ \pm 2 \cdot z^{2k} \end{cases} + \dots$$

If $|a_\ell| > 1$ however, the theorem is already proved; thus we can assume $\ell > 2k$, $a_k = \pm 1$. Applying Lemma 1 to $c_k, c_{2k}$ we have

$$-(1-c^2 - \frac{c_k^2}{1+c}) \le c_{2k} \le 1 - c^2 - \frac{c_k^2}{1-c},$$

$$-(1-c^2 - \frac{d_k^2}{1-c}) < -d_{2k} \le 1 - c^2 - \frac{d_k^2}{1+c}.$$

On adding the inequalities we get

$$-2(1-c^2)+(\frac{c_k^2}{1+c} + \frac{d_k^2}{1-c}) \le c_{2k}-d_{2k} \le 2(1-c^2)-(\frac{c_k^2}{1-c} + \frac{d_k^2}{1+c}).$$

Let $H(x) = \frac{x^2}{1+c} + \frac{(c-x)^2}{1-c}$. Since $|c_k| + |d_k| = c$ it follows that

$$-2(1-c^2)+\min_{x\in I} H(x) \le c_{2k}-d_{2k} \le 2(1-c^2) - \min_{x\in I} H(x),$$

where $I$ is the interval $[c+c^2-1, \ 1-c^2]$.

In view of $c_{2k} - d_{2k} = a_k c_k$ this implies

$$(5) \qquad c^2 + c - 1 \le |c_k| \le 2(1 - c^2) - \min_{x\in I} H(x).$$

Since

$$\frac{1}{2} H'(x) = \frac{x}{1+c} + \frac{x-c}{1-c} = \frac{2x-c-c^2}{1-c^2}$$

$H(x)$ attains its minimum at $x = \frac{1}{2}(c + c^2)$.

If $\frac{c+c^2}{2} \leq 1 - c^2$ we have $3c^2 + c - 2 < 0$, hence

$c < \frac{2}{3}$, $c^{-1} > 1.5 > \theta_o$. If $\frac{c+c^2}{2} > 1 - c^2$ H takes its minimum

in I at $x = 1 - c^2$. By (5)

$$c^2 + c - 1 \leq 2(1-c^2) - \frac{(1-c^2)^2}{1+c} - \frac{(c+c^2-1)^2}{1-c},$$

$$(1-c)(c^2+c-1) \leq 2(1-c^2)(1-c) - (1-c^2)(1-c)^2 - (c+c^2-1)^2$$

and on reduction

$$c^3 + c^2 - 1 \leq 0, \quad c^{-1} \geq \theta_o .$$

This proves the theorem.

<u>Conjecture</u>. For any $P(z) \in \mathbb{Z}[z]$, $P \neq \pm z$, irreducible non-cyclotomic, we have $M(P) \geq \sigma_o = 1.17\ldots$, where $\sigma_o$ is the least known Salem number, satisfying the equation

$$\sigma_o^{10} + \sigma_o^9 - \sigma_o^7 - \sigma_o^6 - \sigma_o^5 - \sigma_o^4 - \sigma_o^3 + \sigma_o + 1 = 0.$$

For reciprocal polynomials, the best published result due to Dobrowolski 1979, is

$$M(P) > 1 + (1 - \varepsilon)\left(\frac{\log \log n}{\log n}\right)^3, \quad n = |P| > n_o(\varepsilon).$$

D.Cantor and E.G.Straus and independently U.Rausch have improved $1 - \varepsilon$ to $2 - \varepsilon$ (unpublished).

Definition 17. *A field $K$ is Kroneckerian if it is totally real or totally complex quadratic extension of a totally real field.*

Corollary. A field $K$ is Kroneckerian if and only if $\bar{K} = K$ and for every isomorphic injection $\sigma$ of $K$ into $\mathbb{C}$ and every $\alpha \in K$ $\overline{\alpha^\sigma} = \bar{\alpha}^\sigma$.

Proof. The necessity of the condition is obvious. In order to prove the sufficiency consider the maximal real subfield $K_0$ of $K$. For $\alpha \in K_0$ $\overline{\alpha^\sigma} = \alpha^\sigma$, hence $K_0$ is totally real Moreover since $\bar{K} = K$ for all $\alpha \in K$ we have $\alpha + \bar{\alpha} \in K_0$ and $\alpha\bar{\alpha} \in K_0$ hence $[K : K_0] \leq 2$; also $\alpha \neq \bar{\alpha}$ implies $\alpha^\sigma \neq \bar{\alpha}^\sigma$. Therefore either $K = K_0$ or $K$ is a totally complex quadratic extension of $K_0$.

Definition 18. *A polynomial $F \in \mathbb{C}[x_1, \ldots, x_s]$ is bireciprocal if*

$$JF(x_1^{-1}, \ldots, x_s^{-1}) = \text{const}\ \bar{F}(x_1, \ldots, x_s).$$

Corollary. $F \in \mathbb{R}[x_1, \ldots, x_s]$ is bireciprocal if and only if it is reciprocal.

Theorem 29 (Schinzel 1975a). *Let $K$ be a Kroneckerian field and $P(z)$ a polynomial over $K$ non-bireciprocal with $P(0) \neq 0$. Then*

$$\prod_{\sigma \in S} M(P^\sigma) \geq \left(\frac{1+\sqrt{5}}{2}\right)^{\frac{\deg K}{2}} N_{K/\mathbb{Q}}(C(P)).$$

*Here $S$ is the set of all isomorphic injections of $K$*

*into* $\mathbb{C}$ *and* $C(P)$ *is the content of* $P$. *The equality sign is possible only if* $\sqrt{5} \in K$.

Lemma. We have the inequalities:

$$(1) \quad \prod_{i=1}^{n} (y_i - 1) \leq ((y_1 \cdots y_n)^{1/n} - 1)^n \quad \text{for} \quad y_i > 1,$$

$$(2) \quad y + \sqrt{c+y^2} \geq (1 + \sqrt{c+1}) y^{1/\sqrt{c+1}}, \quad y > 0,$$

and the equality sign in (1) is attained only if $y_1 = y_2 = \ldots = y_n$.

Proof. The function $F(x) = \log(e^x - 1)$ defined for $x > 0$ is concave, since $F''(x) < 0$. Hence

$$\frac{\log(e^{x_1} - 1) + \ldots + \log(e^{x_n} - 1)}{n} \leq \log(e^{\frac{x_1 + \ldots + x_n}{n}} - 1),$$

with the equality attained only if $x_1 = x_2 = \ldots = x_n$. Replacing $e^{x_i}$ by $y_i$ we get

$$\frac{\log \prod_{i=1}^{n} (y_i - 1)}{n} \leq \log((y_1 \cdots y_n)^{1/n} - 1),$$

and the first inequality follows.

For the second inequality, let $G(x) = \log(e^x + \sqrt{c+e^{2x}})$. The function $G(x)$ is convex, since $G''(x) > 0$. Thus

$$G(x) \geq G(0) + G'(0)x = \log(1 + \sqrt{c+1}) + \frac{1}{\sqrt{c+1}} x;$$

$$e^x + \sqrt{c+e^{2x}} \geq (1 + \sqrt{c+1}) e^{x/\sqrt{c+1}}.$$

Replacing $e^x$ by $y$ we get the lemma.

Proof of Theorem 29. Case 1. $|p_o| = |P(0)|$. Let $Q(z) = z^{|P|}\bar{P}(z^{-1})$ and let us consider

$$\frac{\bar{P}(0)P(z)}{p_oQ(z)} = \frac{f(z)}{g(z)},$$ where $f,g$ are holomorphic in

$|z| < \delta$, $\delta > 1$ and satisfy $|f(z)| = |g(z)| = 1$ on $|z| = 1$ by Lemma 2 to Theorem 28.

We have also

$$\frac{\bar{P}(0)p(z)}{p_oQ(z)} = 1 + a_kz^k + \ldots,$$

where the expansion on the right hand side is infinite, since $P(z)$ is not bireciprocal.

Hence

$$p_oQ(z) + a_kp_oQ(z)z^k + \ldots = \bar{P}(0)P(z)$$

and $a_k|p_o|^2z^k$ is the first term in the Taylor expansion of $\bar{P}(0)P(z) - p_oQ(z)$. All coefficients of $P$ are divisible by $C(P)$, those of $Q$ by $C(\bar{P})$ so

(3) $$\frac{C(P)C(\bar{P})}{(p_o\bar{p}_o)} \mid a_k.$$

If we write

$$f(z) = c + c_1z + \ldots$$

$$g(z) = c + d_1z + \ldots$$

we have $c_k = d_k + a_kc$. By Lemma 1 to Theorem 28, we have also

$$|c_k| \leq 1 - |c|^2, \quad |d_k| \leq 1 - |c|^2, \quad \text{hence}$$

$$|a_k||c| \leq 2(1 - |c|^2).$$

If we replace the polynomial $P$ by $P^{(\sigma)}$ then $Q^\sigma$ is replaced by $Q$ and $a_i$ by $a_i^\sigma$ in virtue of Corollary to Definition 17.

The coefficients of $f,g$ became $c_{i\sigma}, d_{i\sigma}$ with $c_{o\sigma} = d_{o\sigma} = c_\sigma$. Thus we get

$$|a_k^\sigma c_\sigma| \leq 2(1 - |c_\sigma|^2),$$

$$2|c_\sigma|^2 + |a_k||c_\sigma| - 2 \leq 0.$$

By Lemma 2 to Theorem 28 $|c_\sigma| = |p_o^\sigma| M(P^\sigma)^{-1}$, so we are interested in estimating $|c_\sigma^{-1}|$ from below. We get

$$2|c_\sigma|^{-2} - |a_k^\sigma||c_\sigma|^{-1} - 2 \geq 0,$$

or

$$|c_\sigma|^{-1} \geq \frac{|a_k^\sigma| + \sqrt{|a_k^\sigma|^2 + 16}}{4} \geq \left(\frac{1+\sqrt{17}}{4}\right)|a_k^\sigma|^{1/\sqrt{17}} \quad \text{by (2).}$$

Hence

$$\prod_{\sigma \in S} M(p^\sigma) = \prod_{\sigma \in S} |p_o^\sigma c_\sigma^{-1}| \geq \left(\frac{1+\sqrt{17}}{4}\right)^{\deg K} |N_{K/\mathbb{Q}} a_k{}^{1/\sqrt{17}} N_{K/\mathbb{Q}} p_o|$$

$$\geq \left(\frac{1+\sqrt{17}}{4}\right)^{\deg K} \left| N_{K/\mathbb{Q}}\left(\frac{C(P)}{p_o}\right)\right|^{1/\sqrt{17}} |N_{K/\mathbb{Q}} p_o|$$

and we have only to observe

$$\frac{1+\sqrt{17}}{4} > \sqrt{\frac{1+\sqrt{5}}{2}}, \quad \frac{1}{\sqrt{17}} < 1, \quad N_{K/\mathbb{Q}} C(P) \leq |N_{K/\mathbb{Q}} p_o|$$

in order to obtain the assertion.

Case 2. $|p_o| \neq |P(0)|$.

We have $|p_o^2| \neq |P(0)|^2$,

$$(4) \qquad \prod_{\sigma \in S} \left| |p_o^\sigma|^2 - |P^\sigma(0)|^2 \right| = \left| N_{K/\mathbb{Q}}((p_o \bar{p}_o)^2 - (P(0)\bar{P}(0)^2)) \right|$$

$$\geq (N_{K/\mathbb{Q}} C(P))^2.$$

Without loss of generality we may assume that all coefficients of $P$ are algebraic integers. Write $N_o = |N_{K/\mathbb{Q}} p_o|$,

$$N_1 = |N_{K/\mathbb{Q}} P(0)|, \quad N_2 = N_{K/\mathbb{Q}}(C(P)).$$

Define a subset $A$ of $S$ as follows:

$$\sigma \in A \quad \text{if} \quad |P^{(\sigma)}(0)| > |p_o^\sigma|.$$

We have from (4)

$$(5) \qquad E = \prod_{\sigma \in A} |p_o^\sigma|^2 \left| \left|\frac{P^\sigma(0)}{p_o^\sigma}\right|^2 - 1 \right| \cdot \prod_{\sigma \notin A} |P^\sigma(0)|^2 \cdot \left| \left|\frac{p_o^\sigma}{P^\sigma(0)}\right|^2 - 1 \right| \geq N_2^2.$$

We are interested in the product $\Pi$ satisfying

$$(6) \qquad \Pi = \prod_{\sigma\in S} \max(|P^{\sigma}(0)|,|p_0^{\sigma}|) \leq \prod_{\sigma\in S} M(P^{(\sigma)}).$$

Now,

$$E = \frac{N_1^2N_0^2}{\Pi^2} \prod_{\sigma\in A} \left|\left|\frac{P^{\sigma}(0)}{p_0^{\sigma}}\right|^2 - 1\right| \cdot \prod_{\sigma\notin A} \left|\left|\frac{p_0^{\sigma}}{P^{\sigma}(0)}\right|^2 - 1\right|,$$

$$\prod_{\sigma\in A} \left|\frac{P^{\sigma}(0)}{p_0^{\sigma}}\right|^2 \prod_{\sigma\notin A} \left|\frac{p_0^{\sigma}}{P^{\sigma}(0)}\right| = \frac{\Pi^4}{N_0^2N_1^2}.$$

Hence by (1) and (5)

$$N_2^2 \leq \frac{N_0^2N_1^2}{\Pi^2}(\Pi^{4/n}N_0^{-2/n}N_1^{-2/n}-1)^n = (\Pi^{2/n} - \Pi^{-2n}N_0^{2/n}N_1^{2/n})^n ;$$

$$\Pi^{2/n} - \Pi^{-2/n}N_0^{2/n}N_1^{2/n} \geq N_2^{2/n};$$

$$\Pi^{2/n} \geq \frac{N_2^{2/n} + \sqrt{N_2^{4/n}+4(N_0N_1)^{2/n}}}{2} = \frac{\left(\frac{N_2^2}{N_0N_1}\right)^{1/n} + \sqrt{\left(\frac{N_2^2}{N_0N_1}\right)^{2/n}+4}}{2} \cdot (N_0N_1)^{1/n}$$

and by (2) we have

$$\Pi^{2/n} \geq \frac{1+\sqrt{5}}{2}\left(\frac{N_2^2}{N_0N_1}\right)^{1/n\sqrt{5}} \cdot (N_0N_1)^{1/n}.$$

Since $N_0 \geq N_2$, $N_1 \geq N_2$ we get by (6)

$$\prod_{\sigma\in S} M(P^{\sigma}) \geq \Pi \geq \left(\frac{1+\sqrt{5}}{2}\right)^{n/2} N_2^{1/\sqrt{5}} (N_0N_1)^{1/2-1/2\sqrt{5}} \geq \left(\frac{1+\sqrt{5}}{2}\right)^{n/2} N_2,$$

precisely what is asserted.

By the lemma above the equality in the middle can occur only if

$$\frac{|P^{\sigma}(0)|}{|p_0^{\sigma}|} = \begin{cases} \frac{1+\sqrt{5}}{2} & \text{for } \sigma \in A \\ \\ \frac{-1+\sqrt{5}}{2} & \text{for } \sigma \notin A. \end{cases}$$

Since $\frac{|P(0)|^2}{|p_0|^2} \in K$, we get $\sqrt{5} \in K$

and the theorem follows.

Remark. All polynomials $P$ for which there is equality in Theorem 29 have been determined by Bazylewicz 1976.

## Section 21. Application to reducibility of polynomials

Theorem 30. (Vicente Gonçalvez 1956). *If* $F(z) = \prod_{i=1}^{n} (z - \alpha_i) \in \mathbb{C}[z]$, $n \geq 1$, *then*

$$(1) \qquad \prod_{\substack{i=1 \\ |\alpha_i|>1}}^{n} |\alpha_i|^2 + \prod_{\substack{i=1 \\ |\alpha_i|<1}}^{n} |\alpha_i|^2 \leq \|F\|,$$

*where an empty product is meant to be 1.*

We recall that $\|F\|$ is the sum of the absolute values of the coefficients of $F$ squared.

Proof. (Ostrowski 1960). (1) holds for $F = z^n$. Hence dividing if necessary $F$ by a power of $z$ we may assume that $F(0) \neq 0$.

Let $F(z) = G(z)H(z)$, $G(z) = \prod_{\substack{i=1 \\ |\alpha_i|\geq 1}}^{n} (z - \alpha_i)$, $H(z) = \prod_{\substack{i=1 \\ |\alpha_i|<1}}^{n} (z-\alpha_i)$

and compute $F(z)\bar{F}(z^{-1}) = G(z)H(z)\bar{G}(z^{-1})\bar{H}(z^{-1})$

$$= (G(z)\bar{H}(z^{-1}))\cdot(\bar{G}(z^{-1})H(z)).$$

Look at the constant term: on the left it is $\|F\|$, on the right $\|E\|$, where $E(z) = z^{|H|}G(z)\bar{H}(z^{-1}) =$

$$= \prod_{\substack{i=1 \\ |\alpha_i|\geq 1}}^{n} (z - \alpha_i) \prod_{\substack{i=1 \\ |\alpha_i|<1}}^{n} (1 - \bar{\alpha}_i z).$$

The absolute value of the leading coefficient of $E$ is

$\prod_{\substack{i=1 \\ |\alpha_i|<1}}^{n} |\alpha_i|$. The absolute value of its last coefficient is $\prod_{\substack{i=1 \\ |\alpha_i|>1}}^{n} |\alpha_i|$. Hence

$$\|F\| \geq \prod_{\substack{i=1 \\ |\alpha_i|>1}}^{n} |\alpha_i|^2 + \prod_{\substack{i=1 \\ |\alpha_i|<1}}^{n} |\alpha_i|^2 .$$

Corollary 1. (Landau 1905, Specht 1949). For a monic polynomial $F \in \mathbb{C}[z]$, $M(F) \leq \|F\|^{\frac{1}{2}}$.

Corollary 2. The equality sign in (1) is attained only if

$$F(z)\bar{F}(z^{-1}) = \bar{F}(0)z^{|F|} + \|F\| + F(0)z^{-|F|} .$$

Proof. The equality in (1) is attained only if $E$ has just two terms, but then $E(z)\bar{E}(z^{-1}) = F(z)\bar{F}(z^{-1})$ has just three terms the constant one being $\|F\|$.

Theorem 31. (Schinzel 1975a for $s = 1$). *Let* $F \in \mathbb{Z}[x_1,\ldots,x_s]$ *be prime to* $x_1\ldots x_s$ *and* $\Omega = \Omega_2(F,K)$ *be the number of non-bireciprocal factors of* $F$ *irreducible over a Kroneckerian field* $K$, *counted with multiplicities. Then*

$$\left(\frac{1+\sqrt{5}}{2}\right)^{\Omega} + \left(\frac{1+\sqrt{5}}{2}\right)^{-\Omega} \leq \|F\| .$$

*The equality sign is possible only if* $\sqrt{5} \in K$,

$F = JF_0(x_1^{\delta_1} \ldots x_s^{\delta_s})$ *for an* $F_0 \in \mathbf{Z}[z]$ *and some integers* $\delta_1, \ldots, \delta_s$ *and*

$$F_0(z)F_0(z^{-1}) = \pm z^{|F_0|} + \|F\| \pm z^{-|F_0|}.$$

<u>Lemma</u>. If a polynomial $H \in K[x_1, \ldots, x_s]$ prime to $x_1 x_2 \ldots x_s$ is not bireciprocal, but $JH(x^{v_1}, x^{v_2}, \ldots, x^{v_s})$ is bireciprocal then for some integers $\gamma_1, \ldots, \gamma_s$ we have

$$\sum_{i=1}^{s} \gamma_i v_i = 0, \qquad 0 < \max_{1 \le i \le s} |\gamma_i| \le 2|H|. \tag{2}$$

<u>Proof</u>. Let the degree of $H$ with respect to $x_j$ be $h_j$, $\mathbf{h} = \langle h_1, \ldots, h_s \rangle$ and

$$H(x_1, \ldots, x_s) = \sum a_{\boldsymbol{\alpha}} x_1^{\alpha_1} \ldots x_s^{\alpha_s},$$

where the summation is taken over all integral vectors $\boldsymbol{\alpha} = \langle \alpha_1, \ldots, \alpha_s \rangle$ satisfying $0 \le \alpha_j \le h_j$. Clearly

$$J\bar{H}(x_1^{-1}, \ldots, x_s^{-1}) = \sum \bar{a}_{\mathbf{h}-\boldsymbol{\alpha}} x_1^{\alpha_1} \ldots x_s^{\alpha_s}$$

and there exist integral vectors $\boldsymbol{\alpha}_j$ and $\boldsymbol{\alpha}_{-j}$ $(1 \le j \le s)$ such that

$$\alpha_{jj} = h_j, \quad a_{\boldsymbol{\alpha}_j} \ne 0, \quad \alpha_{-jj} = 0, \quad a_{\boldsymbol{\alpha}_{-j}} \ne 0,$$

$\alpha_{\pm jj}$ being the j-th component of $\boldsymbol{\alpha}_{\pm j}$.

Let the product $\boldsymbol{\alpha}\mathbf{v}$ taken over all $\boldsymbol{\alpha}$ with $a_\alpha \neq 0$ attains its minimum for $\alpha = \boldsymbol{\alpha}_m$, maximum for $\alpha = \alpha_n$. We have

$$JH(x^{v_1},\ldots,x^{v_s}) = x^{-\boldsymbol{\alpha}_m \mathbf{v}} \Sigma a_\alpha x^{\alpha\mathbf{v}},$$

$$JH(x^{-v_1},\ldots,x^{-v_s}) = x^{\boldsymbol{\alpha}_n \mathbf{v}} \Sigma a_\alpha x^{-\boldsymbol{\alpha}\mathbf{v}}.$$

All the exponents are different unless (2) holds (even with $|H|$ instead of $2|H|$). In particular, $H(x^{v_1},\ldots,x^{v_s}) \neq 0$. The equality

$$JH(x^{-v_1},\ldots,x^{-v_s}) = cJ\bar{H}(x^{v_1},\ldots,x^{v_s})$$

implies

$$x^{\boldsymbol{\alpha}_n \mathbf{v}} \Sigma a_\alpha x^{-\boldsymbol{\alpha}\mathbf{v}} = cx^{-\boldsymbol{\alpha}_m \mathbf{v}} \Sigma \bar{a}_\alpha x^{\alpha\mathbf{v}}. \tag{3}$$

In particular, we have for each integer $j \neq 0$ with $|j| \leq s$ and a suitable $\boldsymbol{\beta}_j$

$$c\bar{a}_{\boldsymbol{\alpha}_j} x^{(\boldsymbol{\alpha}_j - \boldsymbol{\alpha}_m)\mathbf{v}} = a_{\boldsymbol{\beta}_j} x^{(\boldsymbol{\alpha}_n - \boldsymbol{\beta}_j)\mathbf{v}}.$$

If $\boldsymbol{\alpha}_j + \boldsymbol{\beta}_j - \boldsymbol{\alpha}_m - \boldsymbol{\alpha}_n \neq 0$ we get again (2), otherwise for $j = 1,2,\ldots,s$

$$\alpha_{mj} + \alpha_{nj} = \alpha_{jj} + \beta_{jj} \geq \alpha_{jj} = h_j ,$$

$$\alpha_{mj} + \alpha_{nj} = \alpha_{-jj} + \beta_{-jj} = \beta_{-jj} \leq h_j ,$$

thus $\alpha_{mj} + \alpha_{nj} = h_j$.

Hence $\boldsymbol{\alpha}_m + \boldsymbol{\alpha}_n = \mathbf{h}$ and

$$x^{(\boldsymbol{\alpha}_m + \boldsymbol{\alpha}_n)\mathbf{v}} \Sigma\, a_{\boldsymbol{\alpha}} x^{-\boldsymbol{\alpha}\mathbf{v}} = \Sigma\, a_{\mathbf{h}-\boldsymbol{\alpha}} x^{\boldsymbol{\alpha}\mathbf{v}}.$$

This together with (3) implies

$$a_{\mathbf{h}-\boldsymbol{\alpha}} = c\bar{a}_{\boldsymbol{\alpha}}$$

and $JH(x_1^{-1},\ldots,x_s^{-1}) = c\bar{H}(x_1,\ldots,x_s)$, contrary to the assumption.

Proof of Theorem 31. Consider first the case $s = 1$ and let

$$F = P_o P_1 \ldots P_\Omega ,$$

where $P_o$ is bireciprocal and $P_1,\ldots,P_\Omega$ are non-bireciprocal and irreducible over $K$, not necessarily distinct. Denoting by $S$ the set of all isomorphic injections of $K$ into $\mathbb{C}$ we have for all $\sigma \in S$

$$F = P_o^\sigma P_1^\sigma \ldots P_\Omega^\sigma$$

and since the function $M(P)$ is multiplicative

$$M(F) = \prod_{j=0}^{\Omega} M(P_j^{\sigma}) \geq |p_o^{\sigma}| \prod_{j=1}^{\Omega} M(P_j^{\sigma})$$

where $p_o$ is the leading coefficient of $P_o$.

Hence

$$M(F)^{\delta} \geq |N_{K/\mathbb{Q}} p_o| \prod_{j=1}^{\Omega} \prod_{\sigma \in S} M(P_j^{\sigma}),$$

where $\delta$ equal to the cardinality of $S$ is the degree of K.

Applying Theorem 29 to the inner product we get

$$M(F)^{\delta} \geq |N_{K/\mathbb{Q}} p_o| \cdot \prod_{j=1}^{\Omega} \left(\frac{1+\sqrt{5}}{2}\right)^{\frac{\delta}{2}} N_{K/\mathbb{Q}}\, C(P_j)$$

$$= |N_{K/\mathbb{Q}} p_o| \left(\frac{1+\sqrt{5}}{2}\right)^{\frac{\delta\Omega}{2}} N_{K/\mathbb{Q}} \prod_{j=1}^{\Omega} C(P_j)$$

$$= |N_{K/\mathbb{Q}} p_o| \left(\frac{1+\sqrt{5}}{2}\right)^{\frac{\delta\Omega}{2}} N_{K/\mathbb{Q}}\, C(F/P_o)$$

$$= \left(\frac{1+\sqrt{5}}{2}\right)^{\frac{\delta\Omega}{2}} N_{K/\mathbb{Q}}\, C(F) N_{K/\mathbb{Q}} \frac{(p_o)}{CP_o} \geq \left(\frac{1+\sqrt{5}}{2}\right)^{\frac{\delta\Omega}{2}}$$

with the equality in the first row possible only if $\sqrt{5} \in K$. Taking $\delta$-th roots of both sides we obtain

$$M(F) \geq \left(\frac{1+\sqrt{5}}{2}\right)^{\frac{\Omega}{2}} \tag{4}$$

with the equality possible only if $\sqrt{5} \in K$.

Now let us apply Theorem 30 and its Corollary 2 to the polynomial $f_o^{-1}F$, where $f_o$ is the leading coefficient of F. We get

$$|f_o|^{-2}M(F)^2 + |F(0)|^2M(F)^{-2} \leq f_o^{-2}\|F\|,$$

and

$$M(F)^2 + |f_oF(0)|^2M(F)^{-2} \leq \|F\|$$

with the equality possible only if

$$F(z)\bar{F}(z^{-1}) = f_o\bar{F}(0)z^{|F|} + \|F\| + \bar{f}_oF(0)\bar{z}^{|F|}.$$

Since $F \in \mathbf{Z}[z]$ and so $|f_oF(0)| \geq 1$ we obtain

$$M(F)^2 + M(F)^{-2} \leq \|F\|$$

with the equality possible only if $f_oF(0) = \pm 1$.

Since $x + x^{-1}$ is increasing for $x \geq 1$ it follows form (4) that

$$\left(\frac{1+\sqrt{5}}{2}\right)^{\Omega} + \left(\frac{1+\sqrt{5}}{2}\right)^{-\Omega} \leq \|F\|$$

with the equality possible only if

$$F(z)F(z^{-1}) = \pm z^{|F|} \pm z^{-|F|} + \|F\|$$

and $\sqrt{5} \in K$.

Consider now the case $s > 1$ and let $d > 2|F| > 0$,

$$G = JS_dF = JF(z,z^d,\dots,z^{d^{s-1}}).$$

By the already proved part of the theorem we have

$$(5) \qquad \left(\frac{1+\sqrt{5}}{2}\right)^{\Omega_2} + \left(\frac{1+\sqrt{5}}{2}\right)^{-\Omega_2} \le \|G\| = \|F\|$$

where $\Omega_2 = \Omega_2(G,K)$.

On the other hand, every non-bireciprocal factor $H$ of $F$ in $K$ has the property that $JS_dH$ is not bireciprocal. Indeed, otherwise we would have by the lemma

$$\sum_{i=1}^{s} \gamma_i d^{i-1} = 0, \quad 0 < \max|\gamma_i| \le 2|H| \le 2|F| \le d-1$$

which is clearly impossible. A non-bireciprocal polynomial has a non-bireciprocal irreducible factor, thus

$$\Omega = \Omega_2(F,K) \le \Omega_2(G,K) = \Omega_2$$

and we get from (5)

$$\left(\frac{1+\sqrt{5}}{2}\right)^{\Omega} + \left(\frac{1+\sqrt{5}}{2}\right)^{-\Omega} \le \|F\|$$

with the equality possible only if $\sqrt{5} \in K$ and

$$G(z)G(z^{-1}) = \pm z^{|G|} + \|G\| \pm z^{-|G|}.$$

Since $d > 2|F|$ the substitution $S_d$ transforms not only distinct terms of $F$ into distinct terms of $S_dF$, but distinct terms of $F \cdot F(x_1^{-1},\dots,x_s^{-1})$ into distinct terms of $S_d F \cdot S_d F(x_1^{-1},\dots,x_s^{-1}) = G(z) \cdot G(z^{-1})$. Let the leading term of $F$ in the antilexicographic order be $ax_1^{\alpha_1}\dots x_s^{\alpha_s}$ and the last term $bx_1^{\beta_1}\dots x_s^{\beta_s}$. It follows that

$$(6) \qquad F(x_1,\dots,x_s)F(x_1^{-1},\dots,x_s^{-1})$$

$$= \pm x_1^{\alpha_1-\beta_1}\dots x_s^{\alpha_s-\beta_s} + \|F\| \pm x_1^{\beta_1-\alpha_1}\dots x_s^{\beta_s-\alpha_s}.$$

Let $(\alpha_1 - \beta_1,\dots,\alpha_s - \beta_s) = \gamma,\quad \alpha_i - \beta_i = \gamma\delta_i \quad (1 \le i \le s),$

$$\pm z^{2\gamma} + \|F\|\, z^{\gamma} \pm 1 \overset{\text{can}}{\underset{\mathbb{Q}}{=}} \text{const} \prod_{j=1}^{\ell} F_j(z)^{e_j}.$$

By Lemma 7 to Theorem 27 the polynomials $JF_j(x_1^{\delta_1}\dots x_s^{\delta_s})$ are irreducible over $\mathbb{Q}$, hence

$$F(x_1,\dots,x_s)JF(x_1^{-1},\dots,x_s^{-1}) \overset{\text{can}}{\underset{\mathbb{Q}}{=}} \text{const} \prod_{j=1}^{\ell} JF_j(x_1^{\delta_1}\dots x_s^{\delta_s})^{e_j}$$

and we have for a suitable $F_o \in \mathbf{Z}[z]$

$$F(x_1,\dots,x_k) = JF_o(x_1^{\delta_1}\dots x_s^{\delta_s}).$$

It follows that for each $i \le s$ we have $\alpha_i\beta_i = 0$, $\alpha_i - \beta_i = |F_o|\delta_i$ and by (6)

$$F_o(z)F_o(z^{-1}) = \pm z^{|F_o|} + \|F\| \pm z^{-|F_o|}.$$

Corollary 1. A trinomial $T(x) = x^p + \varepsilon x^q + \eta$ $(\varepsilon, \eta = \pm 1)$, deprived of all its cyclotomic factors, is irreducible over any Kroneckerian field $K$ unless $p = 2q$, $\eta = 1$, or $p = 2q$, $\eta = -1$, $\sqrt{5} \in K$ when

$$T(x) = (x^q - \zeta_3\varepsilon)(x^q - \zeta_3^2\varepsilon)$$

or $$T(x) = (x^q + \frac{\varepsilon+\sqrt{5}}{2})(x^q + \frac{\varepsilon-\sqrt{5}}{2}),$$ respectively.

Proof. We have $\|T\| = 3$ and by the theorem the number $\Omega$ of non-bireciprocal factors of $T$ irreducible over K counted with multiplicities satisfies

$$(\frac{1+\sqrt{5}}{2})^{\Omega} + (\frac{1+\sqrt{5}}{2})^{-\Omega} \leq 3, \quad \text{thus} \quad \Omega \leq 2.$$

If $\Omega = 2$ there is equality, hence $\sqrt{5} \in K$ and

$$(x^p + \varepsilon x^q + \eta)(x^{-p} + \varepsilon x^{-q} + \eta) = \pm x^p + 3 \pm x^{-p},$$

which gives $p = 2q$, $\eta = -1$.

Otherwise $\Omega = 0$ or $1$. If $\Omega = 0$, $T$ is bireciprocal, $p = 2q$, $\eta = 1$. It remains to show that possible bireciprocal factors of $T(x)$ are cyclotomic over $K$.

If $T(x) = T(x^{-1}) = 0$, then

$$x^p + \varepsilon x^q = -\eta,$$

$$x^{-p} + \varepsilon x^{-q} = -\eta,$$

thus $\varepsilon x^{p+q} = 1$, hence $x$ is a root of unity.

Remark 1. $T(x)$ can have cyclotomic factors if

$$\frac{p}{(p,q)} + \frac{q}{(p,q)} \equiv 0 \pmod 3, \qquad \eta = 1.$$

Remark 2. The equality in Theorem 31 is attained for arbitrarily large values of $\Omega$. An example is

$$F(x) = x^{2m} - [(\tfrac{1+\sqrt{5}}{2})^m + (\tfrac{1-\sqrt{5}}{2})^m]\, x^m - 1, \quad m \text{ odd},$$

$$K = \mathbb{Q}(\sqrt{5}, \zeta_m).$$

$F$ factors over $K$ into linear factors:

$$F(x) = (x^m - (\tfrac{1+\sqrt{5}}{2})^m)(x^m - (\tfrac{1-\sqrt{5}}{2})^m)$$

$$= \prod_{j=0}^{m-1} (x - \zeta_m^j \tfrac{1+\sqrt{5}}{2})(x - \zeta_m^j \tfrac{1-\sqrt{5}}{2}),$$

hence $\Omega = 2m$. On the other hand

$$||F|| = 1 + (\tfrac{1+\sqrt{5}}{2})^{2m} - 2 + (\tfrac{1+\sqrt{5}}{2})^{-2m} + 1.$$

This shows that Theorems 29 and 31 are best possible.

Remark 3. In analogy with Theorem 31 Smyth's theorem implies for the number $\Omega = \Omega_2(F)$ of irreducible nonreciprocal factors of F over $\mathbb{Q}$ the inequality

$\theta_o^{2\Omega} + \theta_o^{-2\Omega} \leq \|F\|$, provided $(F, x_1 x_2 \dots x_s) = 1$.

In particular $\Omega_2(F) < \dfrac{\log\|F\|}{2 \log \theta_o} = \dfrac{\log\|F\|}{0,56\dots}$.

On the other hand in the example

$$F_o = x^{12} + x^{10} + x^8 + x^6 + x^4 - 3x^2 + 1 =$$

$$= (x^3 + x^2 - 1)(x^3 - x^2 + 1)(x^3 + x + 1)(x^3 + x - 1)$$

we have $\Omega_2(F_o) = 4 = \dfrac{\log 15}{0.67\dots} = \dfrac{\log\|F_o\|}{0.67\dots}$.

One can give an infinite sequence of polynomials $F_n$ with the same ratio $\dfrac{\log\|F_n\|}{\Omega_2(F_n)} = 0.67\dots$ and with $\|F_n\|$ tending to infinity (see Schinzel 1976).

Remark 4. In a way similar to the proof of Theorem 31 the Conjecture from §19 would lead to the proof of the Conjecture from § 18 in the following stronger form

$$\Omega_1(F) < \frac{\log\|F\|}{2 \log c_o}.$$

A suitable analog of the lemma to Theorem 31 has been given by Montgomery and Schinzel 1977.

Theorem 32. (Dörge 1927). *Let* $\psi(t) = a_{-k}t^k + a_{-k+1}t^{(k-1)} + \ldots$ $+ a_{-1}t + a_0 + a_1 t^{-1} + \ldots$ *be a series convergent for* $t > t_0$ *with complex coefficients. If* $e$ *is a positive integer and* $\varphi(t) = \psi(t^{1/e}) \notin \mathbb{Q}[t]$, *but for a sequence of integers* $t_1 < t_2 < \ldots$ *with* $t_i^{1/e} > t_0$ *we have* $\varphi(t_i) \in \mathbb{Z}$, *then there exist a positive real number* $\lambda$ *and integers* $m$ *and* $i_0$ *such that*

$$t_{i+m} - t_i > t_i^{\lambda} \qquad \text{for} \quad i > i_0 .$$

We need the following lemma.

Lemma. (H.A.Schwarz 1882). Let $\varphi(t)$ be a real function continuously differentiable in the interval $t_i \leq t \leq t_{i+m}$ and let real numbers $t_j$ satisfy $t_i < t_{i+1} < \ldots < t_{i+m}$. Then there exists a $\tau$ such that $t_i < \tau < t_{i+m}$ and

$$(1) \qquad \frac{\varphi^{(m)}(\tau)}{m!} = \frac{1}{V_m} \begin{vmatrix} 1 & t_i & t_i^2 & \ldots & t_i^{m-1} & \varphi(t_i) \\ \vdots & & & & & \vdots \\ 1 & t_{i+m} & t_{i+m}^2 & \ldots & t_{i+m}^{m-1} & \varphi(t_{i+m}) \end{vmatrix} ,$$

$$\text{where} \quad V_m = \begin{vmatrix} 1 & t_i & t_i^2 \ldots t_i^{m-1} & t_i^m \\ \cdot & & & \cdot \\ \cdot & & & \cdot \\ \cdot & & & \cdot \\ 1 & t_{i+m} & t_{i+m}^2 \ldots t_{i+m}^{m-1} & t_{i+m}^m \end{vmatrix} .$$

Proof. (due to Walter Strodt, see Lang 1962, p.148).

Consider

$$F(t) = \begin{vmatrix} 1 & t_i & t_i^2 & \ldots & t_i^{m-1} & \varphi(t_i) \\ \cdot & & & & & \cdot \\ \cdot & & & & & \cdot \\ 1 & t_{i+m-1} & t_{i+m-1}^2 & \ldots & t_{i+m-1}^{m-1} & \varphi(t_{i+m-1}) \\ 1 & t & t^2 & \ldots & t^{m-1} & \varphi(t) \end{vmatrix} .$$

We have

$$F(t) = 0 \quad \text{if} \quad t = t_i, \ t_{i+1}, \ldots, t_{i+m-1}.$$

Now consider

$$G(t) = F(t) - c(t-t_i)(t-t_{i+1})\ldots(t-t_{i+m}),$$

$$\text{where} \quad c = \frac{F(t_{i+m})}{(t_{i+m}-t_i)\ldots(t_{i+m}-t_{i+m-1})} .$$

We have $G(t) = 0$ if $t = t_i, \ldots, t_{i+m}$.

Applying the mean value theorem several times we get

$$G'(t) = 0 \quad \text{for} \quad t = t_1', t_2', \ldots, t_m'; \quad t_{i+j-1} < t_j' < t_{j+1},$$

$$G''(t) = 0 \quad \text{for} \quad t = t_1'', t_2'', \ldots, t_{m-1}''; \quad t_j' < t_j'' < t_{j+1}',$$

$$G^{(m)}(\tau) = 0; \qquad t_i < \tau < t_{i+m}.$$

Now,

$$F^{(m)}(t) = \begin{vmatrix} 1 & t_i & t_i^2 \cdots\cdots t_i^{m-1} & \varphi(t_i) \\ \cdot & & & \cdot \\ \cdot & & & \cdot \\ 1 & t_{i+m-1} & t_{i+m-1}^2 \cdots t_{i+m-1}^{m-1} & \varphi(t_{i+m-1}) \\ 0 & 0 & 0 \cdots\cdots 0 & \varphi^{(m)}(t) \end{vmatrix},$$

$$F^m(\tau) = V_{m-1} \cdot \varphi^{(m)}(\tau), \quad \text{thus} \quad G^{(m)}(\tau) = F^{(m)}(\tau) - cm! = 0$$

implies
$$\varphi^{(m)}(\tau)V_{m-1} = \frac{m!F(t_{i+m})}{(t_{i+m}-t_i)\ldots(t_{i+m}-t_{i+m-1})}.$$

This proves the lemma since $V_{m-1}(t_{i+m}-t_i) \ldots (t_{i+m}-t_{i+m-1}) = V_m$ by the familiar property of Vandermonde determinants.

<u>Proof of Theorem 32</u>. If not all coefficients $a_n$ are real let $\ell$ be the least index such that $\text{Im } a_\ell \neq 0$. For all $t$ large enough we have

$$\left|\text{Im} \sum_{n=-k}^{\ell} a_n t^{-n/e}\right| = \left|\text{Im } a_\ell t^{-\ell/e}\right| > \left|\text{Im} \sum_{n=\ell+1}^{\infty} a_n t^{-n/e}\right|,$$

hence for $i$ large enough $\varphi(t_i) \notin \mathbb{R}$, contrary to the assumption. Thus all coefficients of $\psi$ are real and $\varphi$ satisfies the conditions of the lemma. If $\varphi \in \mathbb{R}[t]$, but $\varphi \notin \mathbb{Q}[t]$ let $\ell$ be the least index such that $a_\ell \notin \mathbb{Q}$. Clearly $\ell \leq 0$ and $e|\ell$, moreover $a_n = 0$ if $e \nmid n$. The function $\varphi(t) - \sum_{n=-k}^{\ell-1} a_n t^{-n/e}$ takes rational values for $t = t_1, t_2, \ldots$ . Application of the lemma to this function with $m = \ell/e$ gives an equality with $a_\ell$ on the left hand side and a rational number on the right, a contradiction. It remains to consider the case, where $a_n \in \mathbb{R}$ for all $n \geq -k$ but $\varphi \notin \mathbb{R}[t]$.

Choose $m$ so large that $\varphi^{(m)}(t)$ contains only terms with negative exponents. (Thus $m > k/e$).

For $i > i_o$, $t \geq t_i$ and a suitable $c$ we have

$c^{-1}\ t^{k/e-m} > |\varphi^{(m)}(t)| > 0.$

The determinant in the numerator of the right hand side of (1) is an integer, thus for $i > i_o$ the absolute value of the determinant is at least 1 and

$$\left|\frac{\varphi^{(m)}(\tau)}{m!}\right| \geq \frac{1}{|V_m|} \ .$$

Hence for $i > i_0$

$$|V_m| \geq \frac{m!}{|\varphi^{(m)}(\tau)|} \geq \frac{m!c}{\tau^{\frac{k}{e}-m}} = m!\, c \cdot \tau^{m-\frac{k}{e}} \geq m!\, c\, t_i^{m-\frac{k}{e}}.$$

Each difference in the familiar formula for $V_m$ is in absolute value at most $t_{m+1} - t_i$, thus

$$(t_{m+i}-t_i)^{\frac{m(m-1)}{2}} \geq |V_m| \geq m!c\, t_i^{m-\frac{k}{e}}$$

and the theorem follows for every $\lambda < \frac{2(m-k/e)}{m(m-1)}$.

Corollary. Let $\varphi$ have the meaning of Theorem 32. The number of positive integers $t \leq T$ such that $\varphi(t) \in \mathbf{Z}$ is $O(T^\beta)$, where $\beta < 1$.

Proof. Take $\beta = \frac{1}{\lambda+1}$, where $\lambda$ has meaning of the theorem. Let us order the integers $t$ such that $\varphi(t) \in \mathbf{Z}$ contained in the interval $(T^\beta, T]$ in blocks of $m$ integers:

$$t_i < \ldots < t_{i+m-1} < t_{i+m} < \ldots < t_{i+2m-1} < \ldots .$$

By virtue of the theorem

$$t_{i+m} - t_i > t_i^\lambda > T^{\lambda\beta}.$$

Thus the number B of blocks (including an incomplete one) does not exceed $\frac{T}{T^{\lambda\beta}} + 1$. The total number of integers $t_j \leq T$ for which $\varphi(t_j) \in \mathbb{Z}$ does not exceed

$$i-1+mB \leq T^{\beta}+m\cdot T^{1-\lambda\beta}+m = (m+1)T^{\frac{1}{\lambda+1}} + m.$$

This proves the corollary.

Theorem 33. (Hilbert 1892). *Let* $F_i(x_1,\dots,x_s,t_1,\dots,t_r)$ $(1 \leq i \leq h)$ *be irreducible in an algebraic number field* $K$ *as polynomials in* $s+r$ *variables and let* $P \in K[t_1,\dots,t_r]$, $P \neq 0$. *Then there exist infinitely many integral vectors* $\langle t_1^*,\dots,t_r^*\rangle \in \mathbb{Z}^r$ *such that* $F_i(x_1,\dots,x_s,t_1^*,\dots,t_r^*)$ *are irreducible over* $K$ *for all* $i \leq h$ *and* $P(t_1^*,\dots,t_r^*) \neq 0$.

In the proof we shall need

Theorem F. (Puiseux 1850). *Let* $F(x,t) \in \mathbb{C}[x,t]$. *For* $|t|$ *sufficiently large all solutions of the equation* $F(x,t) = 0$ *are given by* $x = \varphi_l(t)$, *where* $\varphi_l$ *is of the type described in Theorem 32,* $l = 1,\dots,|F|_x$.

Proof. see Bliss 1933, Theorem 13,1.

Bliss assumes (l.c.p.24) that $F(x,t)$ has no factor depending only on $t$ and that its discriminant with respect to $x$ is non-zero. However he observes himself (l.c.p.25) how to get rid of the second assumption and the first one is irrelevant for sufficiently large $t$.

<u>Lemma 1</u>. Let $F \in \mathbb{Z}[x,t]$ be monic with respect to $x$. If $F(x,t) = 0$ has no solution $x \in \mathbb{Q}(t)$ then there exist infinitely many integers $t^*$ such that $F(x,t^*) = 0$ has no solution $x \in \mathbb{Q}$.

<u>Proof</u>. For every $t^* \in \mathbb{Z}$ the polynomial $F(x,t^*)$ is monic with integral coefficients, thus all its rational zeros are integers. By Theorem F all solutions of $F(x,t) = 0$ for sufficiently large $t$ are given by $n = |F|_x$ expansions of the form

$$x(t) = a_{-k}t^{k/e} + a_{-k+1}t^{(k-1)/e} + \ldots .$$

By Corollary to Theorem 32 for each expansion there are only $O(T^\beta)$ positive integers $t \leq T$ such that $x(t)$ is an integer, $\beta$ being a constant less than 1. Since $n \cdot O(T^\beta) = O(T^\beta)$ there exist infinitely many positive integers $t^*$ for which none of $x(t^*)$ is an integer, hence for these $t^*$ no zero of $F(x,t^*)$ is rational.

Lemma 2. Let $K, L$ be fields, $K \subset L$, $[L : K] < \infty$. If $G(\mathbf{x}) \in L[\mathbf{x}]$ is irreducible over $L$ then

$$N_{L/K} G(\mathbf{x}) = c\, F(\mathbf{x})^m,$$

where $F \in K[\mathbf{x}]$ is irreducible over $K$ and is divisible by $G$ in $L[\mathbf{x}]$, $c \in K$.

Proof. Let $F$ be a polynomial in $K[\mathbf{x}]$ irreducible over $K$ and divisible by $G$ in $L[\mathbf{x}]$. Then $G^{(\sigma)}(\mathbf{x}) \mid F(\mathbf{x})$ in $\hat{L}[x]$ for every conjugate $G^{(\sigma)}(\mathbf{x})$ of $G(\mathbf{x})$ over $K(\mathbf{x})$. Hence

$$N_{L/K} G(\mathbf{x}) = \Pi G^{(\sigma)}(\mathbf{x}) \mid F(\mathbf{x})^{[L:K]}.$$

Since $N_{L/K} G(\mathbf{x}) \in K[\mathbf{x}]$ and all divisors of $F(\mathbf{x})^n$ in $K[\mathbf{x}]$ are of the form $cF(\mathbf{x})^m$ the lemma follows.

Lemma 3. Let $F(x_1, \ldots, x_s, t_1, \ldots, t_r)$ be of degree $< d$ in each $x_j$ and irreducible over $K$. There exists a non-zero polynomial $E \in K[t_1, \ldots, t_r]$ such that if $S_d$ is the Kronecker substitution (see § 7), and

$$S_d F = \prod_{\nu=1}^{n} G_\nu(x, t_1, \ldots t_r),$$

where $G_\nu$ are irreducible over $K$ and if for some $t_1^*,\ldots,t_r^*$ in $K$ we have $E(t_1^*,\ldots,t_r^*) \neq 0$ and all the polynomials $G_\nu(x,t_1^*,\ldots,t_r^*)$ are irreducible over $K$, then $F(x_1,\ldots,x_s,t_1^*,\ldots,t_r^*)$ is irreducible over $K$.

Proof. Let $\mathbf{x} = \langle x_1,\ldots,x_s\rangle$, $\mathbf{t} = \langle t_1,\ldots,t_r\rangle$. For every subset $A$ of $\{1,2,\ldots,n\}$ the polynomials $\prod_{\nu\in A} G_\nu(x,\mathbf{t})$ and $\prod_{\nu\notin A} G_\nu(x,\mathbf{t})$ are of degree less than $d^s$ in $x$ hence by Corollary 1 to Definition 7 there exist unique polynomials $G_A(\mathbf{x},\mathbf{t})$ and $H_A(\mathbf{x},\mathbf{t})$ in $K[\mathbf{x},\mathbf{t}]$ satisfying

$$|G_A|_{\mathbf{x}} = \max_{1\le q\le s} |G_A|_{x_q} < d, \quad S_d G_A = \prod_{\nu\in A} G_\nu(x,\mathbf{t}),$$

$$|H_A|_{\mathbf{x}} = \max_{1\le q\le s} |H_A|_{x_q} < d, \quad S_d H_A = \prod_{\nu\notin A} G_\nu(x,\mathbf{t}).$$

If $A$ is a non-empty proper subset of $\{1,2,\ldots,n\}$ we have $G_A \neq \text{const}$, $H_A \neq \text{const}$, hence $G_A H_A \neq F$, since $F$ is irreducible.

Let $E_A(\mathbf{t})$ be the leading coefficient of $G_A H_A - F$ viewed as polynomial in $\mathbf{x}$.

Put

$$E(\mathbf{t}) = \prod_{\substack{A \subsetneq \{1,2,\ldots,n\} \\ A \neq \emptyset}} E_A(\mathbf{t})$$

and take any $\mathbf{t}^*$ such that $E(\mathbf{t}^*) \neq 0$ and such that $G_\nu(x,\mathbf{t}^*)$ are irreducible over $K$ for all $\nu \leq n$.

In order to use Corollaries to Definition 7 consider a factorization

$$S_d F(\mathbf{x},\mathbf{t}^*) = S_d G(\mathbf{x}) S_d H(\mathbf{x}), \quad \text{where} \quad G,H \in K[\mathbf{x}],$$
$$0 < |G| < d, \quad 0 < |H| < d.$$

Since $G_\nu(x,\mathbf{t}^*)$ are all irreducible over $K$ there exists a non-empty proper subset $A$ of $\{1,2,\ldots,n\}$ and a constant $c \neq 0$ such that

$$S_d G = c \prod_{\nu \in A} G_\nu(x,\mathbf{t}^*) = S_d c\, G_A(\mathbf{x},\mathbf{t}^*),$$

$$S_d H = c^{-1} \prod_{\nu \notin A} G_\nu(x,\mathbf{t}^*) = S_d c^{-1} H_A(\mathbf{x},\mathbf{t}^*).$$

By Corollary 1 to Definition 7 we have

$$G = cG_A(\mathbf{x},\mathbf{t}^*), \quad H = c^{-1}H_A(\mathbf{x},\mathbf{t}^*),$$

and by the choice of $E$ : $G(\mathbf{x})H(\mathbf{x}) \neq F(\mathbf{x},\mathbf{t}^*)$, which proves in virtue of Corollary 2 to Definition 7 that $F(\mathbf{x},\mathbf{t}^*)$ is irreducible.

Proof of Theorem 33. Let us consider polynomials $F_i \in K[x_1,\dots,x_s,t_1,\dots,t_r]$ $(i \le h)$ irreducible in $s+r$ variables and a polynomial $P(t_1,\dots,t_r) \ne 0$. The proof will be in several steps.

Step 1. $r = s = 1$, $K = \mathbb{Q}$.

Let $F_i(x,t)$ viewed as a polynomial in $x$ have degree $n_i$, the leading coefficient $a_i(t)$ and satisfy $m F_i \in \mathbb{Z}[x,t]$ for an integer $m \ne 0$ $(1 \le i \le h)$. We have

$$m^{n_i} a_i(t)^{n_i - 1} F_i(x,t) = \tilde{F}_i(m \cdot a_i(t)x,t),$$

where $\tilde{F}_i(x,t) \in \mathbb{Z}[x,t]$ is monic in $x$ and irreducible over $\mathbb{Q}$. Moreover

$$\tilde{P}(t) = P(t) \prod_{i=1}^{j} a_i(t) \ne 0.$$

If for a $t^* \in \mathbb{Z}$ the polynomial $\tilde{F}_i(x,t^*)$ is irreducible over $\mathbb{Q}$ and $\tilde{P}(t^*) \ne 0$ then $F_i(x,t^*)$ is also irreducible over $\mathbb{Q}$ and $P(t^*) \ne 0$. Therefore, replacing if necessary $F_i$ by $\tilde{F}_i$ and $P$ by $\tilde{P}$ we may assume that the $F_i(x,t)$ are monic in $x$ and have integral coefficients for all $i \leq h$.

By Theorem 19 for all positive integers $m,n$ with $m \leq n$ there exists an integral polynomial $\Phi_{mn}(z,u_1,\ldots,u_m,c_1,\ldots,c_n)$ monic in $z$ with the following properties valid for every field $k$ and arbitrary elements $c_1^*,\ldots,c_n^*$ of $k$.

(i). If $u_1^*,\ldots,u_n^* \in k$ and $\mathrm{disc}_z \Phi_{mn}(z,u_1^*,\ldots,u_m^*,c_1^*,\ldots,c_n^*) \neq 0$ then $x^n + c_1^* x^{n-1} + \ldots + c_n^*$ has a factor of degree $m$ in $k[x]$ if and only if $\Phi_{mn}(z,u_1^*,\ldots,u_m^*,c_1^*,\ldots,c_n^*)$ has a zero in $k$.

(ii). If $\mathrm{disc}(x^n + c_1^* x^{n-1} + \ldots + c_n^*) \neq 0$ then every infinite subset of $k$ contains $u_1^*,\ldots,u_m^*$ such that $\mathrm{disc}_z \Phi_{mn}(z,u_1^*,\ldots,u_m^*,c_1^*,\ldots,c_n^*) \neq 0$.

Take $k = \mathbb{Q}(t)$, $F_i(x,t) = x^{n_i} + c_{1i} x^{n_i - 1} + \ldots + c_{n_i i} \in k[x]$. Since $F_i$ is irreducible we have $\mathrm{disc}_x F_i \neq 0$ and by (ii) for each $m < n$ there exist $u_{\ell im} \in \mathbb{Z}$ $(i \leq h,\ \ell \leq m)$ such that

$$\mathrm{disc}_z \Phi_{mn_i}(z,u_{1im},\ldots,u_{mim},c_{1i},\ldots,c_{n_i i}) = D_{im}(t) \neq 0.$$

Consider

$$F(z,t) = \prod_{i=1}^{h} \prod_{m=1}^{n_i-1} \Phi_{mn_i}(z,u_{1im},\ldots,u_{mim},c_{1i},\ldots,c_{n_i i}) \in \mathbf{Z}[z,t].$$

$F(z,t)$ is monic in $z$ and the equation $F(z,t) = 0$ has no solution $z \in \mathbf{Q}(t)$. Otherwise it would follows that for an $i \leq h$ and an $m < n_i$

$\Phi_{mn_i}(z,u_{1im},\ldots,u_{mim},c_{1i},\ldots,c_{n_i i})$ has a zero in $\mathbf{Q}(t)$ and by (i) above $F_i(x,t)$ would be reducible over $k$ contrary to the assumption. Applying Lemma 1 we infer that there exist infinitely many $t^* \in \mathbf{Z}$ such that $F(z,t^*) = 0$ has no solution $z \in \mathbf{Q}$. Among them there are infinitely many $t^*$ such that

$$P(t^*) \prod_{i=1}^{h} \prod_{m=1}^{n_i-1} D_{im}(t^*) \neq 0.$$

It follows from (i) above with $k = \mathbf{Q}$ that for every such $t^*$ all polynomials $F_i(x,t^*)$ are irreducible over $\mathbf{Q}$ $(i \leq h)$.

<u>Step 2</u>. $r = s = 1$, $K$ normal over $\mathbf{Q}$.

As before we may assume that $F_i$ are monic in $x$ for all $i \leq h$. Let $F_i(x,t) = x^{n_i} + b_i(t)x^{n_i-1} + \ldots$ and let

$K = \mathbb{Q}(\Theta)$. We can choose a $c_i \in \mathbb{Q}$ such that for all isomorphisms $\sigma$ of $K$ except the identity

$$c_i \neq \frac{b_i(t)-b_i^{(\sigma)}(t)}{n_i(\Theta^{(\sigma)}-\Theta)} .$$

Then we have $n_i c_i \Theta + b_i(t) \neq n_i c_i \Theta^{(\sigma)} + b_i^{(\sigma)}(t)$ and since $n_i c_i \Theta + b_i(t)$ is the coefficient of $x^{n_i-1}$ in $F_i(x+c_i\Theta,t)$ it follows that the coefficients of $F_i(x+c_i\Theta,t)$ as a polynomial in $x$ and $t$ generate $K$. By Lemma 2 to Theorem 24 $N_{K/\mathbb{Q}}F_i(x+c_i\Theta,t)$ is irreducible over $\mathbb{Q}$. By the already proved case 1 of our theorem there exist infinitely many integers $t^*$ such that all polynomials $N_{K/\mathbb{Q}}F_i(x+c_i\Theta,t^*)$ $(i \leq h)$ are irreducible over $\mathbb{Q}$ and $N_{K/\mathbb{Q}}P(t^*) \neq 0$. Then $F_i(x+c_i\Theta,t^*)$ are irreducible over $K$ and clearly the same holds for $F_i(x,t^*)$ $(i \leq h)$. Moreover $P(t^*) \neq 0$.

<u>Step 3</u>. $r = s = 1$, $K$ arbitrary algebraic number field. Let $D_i(t)$ be the discriminant of $F_i(x,t)$ with respect to $x, L$ be the normal closure of $K$ and $G_i(x,t)$ be a factor of $F_i(x,t)$ monic in $x$ and irreducible over $L$. Since $F_i(x,t)$ is irreducible over $K$ we have by Lemma 2

(2) $$N_{L/K}G_i(x,t) = c_oF_i(x,t)^{m_o}.$$

By the already proved case 2 of our theorem there exist infinitely many integers $t^*$ such that $G_i(x,t^*)$ are irreducible over $L$ $(i \leq h)$ and $P(t^*) \prod_{i=1}^{h} D_i(t^*) \neq 0$. Let $H(x) \in K[x]$ be irreducible over $K$ and divisible by $G_i(x,t^*)$. We have by Lemma 2

$$N_{L/K}G_i(x,t^*) = cH(x)^m$$

and comparison with (2) gives $c_oF_i(x,t^*)^{m_o} = cH(x)^m$. Hence $m_o | m$ and either $F_i(x,t^*) = H(x)$ is irreducible over $K$ or $F_i(x,t^*)$ has a multiple zero. However the latter case is excluded by the condition $D_i(t^*) \neq 0$.

Step 4. $s > 1$, $r = 1$.

Make the Kronecker substitution $S_d$, where $d > \max_{i \leq h} |F_i|_{\mathbf{x}}$ and let

$$S_dF_i = \prod_{\nu=1}^{n_i} G_{\nu i}(x,t)$$

be a factorization of $F_i$ into factors irreducible over $K$. By case 3 we can choose $t^* \in \mathbf{Z}$ in infinitely many ways so that $G_{\nu i}(x,t^*)$ are all irreducible over $K$ for $i \le h, \nu \le n_i$. By Lemma 3 for almost all such $t^*$'s the polynomials $F_i(x_1,\ldots,x_s,t^*)$ are irreducible over $K$ $(i \le h)$ and $P(t^*) \ne 0$.

Step 5. $s > 1$, $r > 1$.

We proceed by induction on $r$. Assume that the theorem holds for $r$ parameters and consider

$F_i(x_1,\ldots,x_s,t_1,\ldots,t_{r+1})$ $(i \le h)$ irreducible over $K$ and $P(t_1,\ldots,t_{r+1}) \ne 0$.

By the already proved case 3 of our theorem we can choose $t^*_{r+1} \in \mathbf{Z}$ so that all polynomials

$F_i(x_1,\ldots,x_s,t_1,\ldots,t_r,t^*_{r+1})$ are irreducible over $K$

and $P_o(t^*_{r+1}) \ne 0$, $P_o$ being the leading coefficient of $P$ viewed as polynomial in $t_1,\ldots,t_r$. By the inductive assumption we can choose $(t^*_1,\ldots,t^*_r) \in \mathbf{Z}^r$ in infinitely many ways so that $F_i(x_1,\ldots,x_s,t^*_1,\ldots,t^*_r,t^*_{r+1})$ are all irreducible over K and $P(t^*_1,\ldots,t^*_{r+1}) \ne 0$.

Remark 1. The proof given above is similar to that of Franz 1931. The original proof by Hilbert 1892 contains a mistake repeated by Lang 1962. Both authors assert that every finite extension $E$ of $\mathbb{Q}$ has the property formulated by Lang as Proposition 3 on p.151 of his book with an extra generality as follows.

Let $E$ be a finite separable extension of a field $k$. Let $f(t,X) \in E(t)[X]$ be irreducible over $E(t)$. Assume that the leading coefficient of $f$ in $X$ is 1, and that if $\sigma$ ranges over the distinct isomorphisms of $E$ over $k$, then the conjugates $f^{\sigma}$ of $f$ are distinct. Put

$$F(t,X) = \Pi f^{\sigma}(t,X).$$

Then $F(t,X)$ is in $k(t)[X]$ and is irreducible.

The above proposition is false already for $k = \mathbb{Q}$, $E = \mathbb{Q}(\sqrt[3]{2})$, $f(x) = x^2 + \sqrt[3]{2}x + \sqrt[3]{4}$ as it has been pointed out in connection with Lemma 2 to Theorem 24.

Remark 2. Let us call a positive integer $t^*$ exceptional if under the assumptions of the theorem $(r = 1)$ at least one of the polynomials $F_i(x_1,\dots,x_s,t^*)$ is reducible over $K$. The above proof easily furnishes an estimate for the number $N(T)$ of exceptional integers $t^* \le T$. Indeed, let $S(F)$ be the set of positive integers $t^*$

for which the equation $F(x,t^*) = 0$ has a solution $x \in \mathbb{Q}$. The proof of Lemma 1 shows that under the assumptions of this lemma the number of elements of $S(F)$ that are less than $T$ is $O(T^\beta)$ with $\beta = \beta(F) < 1$. An analysis of the steps 1-4 shows the existence of a finite set $\Phi$ of polynomials satisfying the assumptions of Lemma 1 such that all exceptional integers belong to the union of sets $S(F)$ $(F \in \Phi)$. Hence

$$N(T) = O(T^\alpha), \qquad \alpha = \max_{F \in \Phi} \beta(F) < 1.$$

Using Theorem E of § 6 one can prove $\beta(F) \leq \frac{1}{2}$ for every $F$ satisfying the assumptions of Lemma 1. Then the above argument gives $N(T) = O(T^{1/2})$. This estimate is best possible, as the example $F_1(x,t) = x^2 - t$ shows. In this connection see Sprindžuk 1979.

In the next section we shall give as Theorem 35 a different refinement of Theorem 33.

Theorem 34. (Davenport, Lewis, Schinzel 1964 for $K = \mathbb{Q}$, $r = 1$, Schinzel 1965 for $K = \mathbb{Q}$, $r \geq 1$). *Let $F(x, t_1, \cdots, t_r)$ be a polynomial over an algebraic number field $K$. If for every $r$ arithmetic progressions $P_1, \ldots, P_r$ in $\mathbb{Z}$ there exist integers $t_l^* \in P_l$ $(l \leq r)$ and an $x^* \in K$ such that $F(x^*, t_1^*, \ldots, t_r^*) = 0$ then there exists a rational function $x(t_1, \ldots, t_r)$ over $K$ such that*

$$(1) \qquad F(x(t_1, \cdots, t_r), t_1, \cdots, t_r) = 0 .$$

Lemma. (Hasse 1932). If $G(x)$ is irreducible over $K$ and $|G| > 1$ then there exist infinitely many prime ideals $\mathfrak{p}$ of $K$ such that $G(x) \equiv 0 \bmod \mathfrak{p}$ is unsolvable in $K$.

Proof. By the Frobenius density theorem, the density of prime ideals $\mathfrak{p}$ of $K$ such that $G(x) \bmod \mathfrak{p}$ factors in a given way is the same as the relative density of permutations in the Galois group $\mathfrak{G}$ of $G$ over $K$ which factor in the same way into cycles (cf Mann 1955, §16). Since $\mathfrak{G}$ is transitive the index of each stability subgroup of $\mathfrak{G}$ is equal to the degree of $\mathfrak{G}$ (cf van der Waerden 1970, §7.9). Since these subgroups are not disjoint there exists an element of $\mathfrak{G}$ not contained in their union, thus with no cycle of length 1 in its factorization. Hence the density of prime ideals satisfying the lemma is positive.

Remark. For $K = \mathbb{Q}$ the lemma has been stated without proof already by Polya and Szegö 1925 as a remark to their solution of Problem 100 in Section VIII. Mann 1955 gives the above proof without a reference to Hasse.

Proof of Theorem 34. Let $\langle t_1, \cdots, t_r \rangle = \mathbf{t}$

$$F(x,\mathbf{t}) \underset{K[\mathbf{t}]}{\overset{can}{=}} F_o(\mathbf{t}) \prod_{i=1}^{j} F_i(x,\mathbf{t})^{e_i} . \tag{2}$$

Replacing $F$ if necessary by a suitable constant multiple, we may assume that $F_i$ have coefficients integral in $K$. Let $a_i(\mathbf{t})$ be the leading coefficient of $F_i$ with respect to $x$.

If for an $i > 0$ we have $|F_i|_x = 1$ then $x(\mathbf{t}) = -F_i(0,\mathbf{t})/a_i(\mathbf{t})$ is the desired solution of (1).

Therefore, assume that $|F_i|_x > 1$ for all $i > 0$. By Theorem 33 there exist integers $\tau_\ell$ such that for all $i \le h$ the polynomials $F_i(x, \tau_1, \cdots, \tau_r)$ are irreducible over $K$ and moreover

$$F_o(\tau_1, \cdots, \tau_r) \prod_{i=1}^{h} a_i(\tau_1, \cdots, \tau_r) \neq 0 .$$

By the lemma for each $i$ there exist infinitely many prime ideals $\mathfrak{p}_i$ such that the congruence

$$F_i(x, \tau_1, \cdots, \tau_r) \equiv 0 (\mathrm{mod}\ \mathfrak{p}_i)$$

has no solution $x \in K$. We can choose $\mathfrak{p}_i$ so that

$$\mathfrak{p}_o \nmid F_o(\tau_1, \cdots, \tau_r) , \mathfrak{p}_i \nmid a_i(\tau_1, \cdots, \tau_r) .$$

Take $P_\ell$ to be the progression $p_o p_1 \cdots p_h t + \tau_\ell (1 \le \ell \le r)$, where $p_i$ is the rational prime divisible by $\mathfrak{p}_i (0 \le i \le h)$ and suppose that

$F(x^*, t^*) = 0$ for a $\mathbf{t}^* \in P_1 \times P_2 \times \cdots \times P_r$ and an $x^* \in K$.

By (2) either $F_o(\mathbf{t}^*) = 0$ or $F_i(x^*, \mathbf{t}^*) = 0$ for an $i \leq h$. However since $F_o$ and $a_i$ have coefficients integral in $K$

$$F(\mathbf{t}^*) \equiv F_o(\tau_1, \cdots, \tau_r) \not\equiv 0 \bmod \mathfrak{p}_o$$

and

$$a_i(\mathbf{t}^*) \equiv a_i(\tau_1, \cdots, \tau_r) \not\equiv 0 \bmod \mathfrak{p}_i (1 \leq i \leq j) .$$

Thus $x^*$ contains $\mathfrak{p}_i$ in a nonnegative power and

$$F_i(x^*, \mathbf{t}^*) \equiv F_i(x^*, \tau_1, \cdots, \tau_r) \not\equiv 0 \bmod \mathfrak{p}_i .$$

The obtained contradiction completes the proof.

Theorem 35. (Schinzel 1965 for $K = \mathbb{Q}$). *Under the assumptions of Theorem 33 there exist $r$ arithmetic progressions $P_1, \cdots, P_r$ in $\mathbb{Z}$ such that if $t_\ell^* \in P_\ell$ for all $\ell \leq r$ then $F_i(x_1, \cdots, x_s, t_1^*, \cdots, t_r^*)$ are all irreducible over $K (i \leq h)$ and $P(t_1^*, \cdots, t_r^*) \neq 0$.*

Proof. Repeating the argument in Step 1 of the proof of Theorem 33 with $k = K(t_1, \ldots, t_r)$ and with Lemma 1 of § 22 replaced by Theorem 34 above we get the case $s = 1$. Repeating the argument in Step 4 with $t$ replaced by $\mathbf{t} = \langle t_1, \ldots, t_r \rangle$ we get the general case.

Remark. Another proof of Theorem 35 in the case $K = \mathbb{Q}$, $h = r = s = 1$ has been given by M.Fried 1974 b. It uses a different approach to the problem indicated in case of Theorem 33 by Eichler 1939.

In his paper Fried has also studied in great depth the special case $K = \mathbb{Q}$, $h = r = s = 1$, $F(x,t) = G(x) - t$.

In the following theorem we both assume more and assert more than in Theorem 34.

Theorem 36 (Schinzel 1965 for $K = \mathbb{Q}$). *Let $K$ be an algebraic number field and $F \in K[x, t_1, \ldots, t_r]$. If for every $r$ arithmetic progressions $P_1, \ldots, P_r$ in $\mathbb{Z}$ there exist integers $t_l^* \in P_l$ $(l \le r)$ and an integer $x^*$ of $K$ such that $F(x^*, t_1^*, \ldots, t_r^*) = 0$ then there exists a polynomial $X(t_1, \ldots, t_r)$ over $K$ such that $F(X(t_1, \ldots, t_r), t_1, \ldots, t_r) = 0$ identically.*

Lemma. Let $\varphi_j(t_1, \ldots, t_r)$ $(j \le h)$ be rational functions, but not polynomials over K. There exist arithmetic progressions $P_1, \ldots, P_r$ in $\mathbb{Z}$ such that if $t_\ell^* \in P_\ell$ $(\ell \le r)$ then no $\varphi_j(t_1, \ldots, t_r)$ $(j \le h)$ is an integer of K.

Proof. (by induction on h). Let $h = 1$ and let $\langle t_1, \ldots, t_r \rangle = \mathbf{t}$. Then $\varphi_1(\mathbf{t}) = \frac{G(\mathbf{t})}{H(\mathbf{t})}$, with $G, H \in K[\mathbf{t}]$, G,H with integral coefficients and $(G,H) = 1$. After relabeling we can assume $|H|_{t_1} > 0$. We have for some $A, B \in K[\mathbf{t}]$ with coefficients integral in K

$$A(\mathbf{t})G(\mathbf{t}) + B(\mathbf{t})H(\mathbf{t}) = R(t_2, \ldots, t_r) \ne 0,$$

R being the resultant of G,H with respect to $t_1$. There exist some $\tau_2, \ldots, \tau_r \in \mathbb{Z}$ such that $R(\tau_2, \ldots, \tau_r) \ne 0$ and the leading coefficient $a_0(\tau_2, \ldots, \tau_r)$ of H with respect to $t_1$ is not zero. Since $N_{K/\mathbb{Q}} H(\tau, \tau_2, \ldots, \tau_r) \in \mathbb{Q}[\tau] \setminus \mathbb{Q}$ we can choose $\tau_1$ so that $c = |N_{K/\mathbb{Q}} H(\tau_1, \tau_2, \ldots, \tau_r)| > |N_{K/\mathbb{Q}} R(\tau_2, \ldots, \tau_r)|$. Take the following arithmetic progressions $P_\ell : ct + \tau_\ell$ $(\ell = 1, \ldots, r)$.

If $\mathbf{t}^* = \langle t_1^*, \ldots, t_r^* \rangle \in P_1 \times P_2 \times \ldots \times P_r$ we have

$$N_{K/\mathbb{Q}}H(\mathbf{t}^*) \equiv N_{K/\mathbb{Q}}H(\tau_1,\ldots,\tau_r) \equiv 0 \pmod{c}.$$

On the other hand

$$N_{K/\mathbb{Q}}R(t_2^*,\ldots,t_r^*) \equiv N_{K/\mathbb{Q}}R(\tau_2,\ldots,\tau_r) \not\equiv 0 \pmod{c}$$

thus $H(\mathbf{t}^*) \nmid R(t_2^*,\ldots,t_r^*)$, $H(\mathbf{t}^*) \nmid G(\mathbf{t}^*)$ and $\varphi_1(\mathbf{t}^*)$ is not an integer of $K$, for $\mathbf{t}^* \in P_1 \times \ldots \times P_r$.

Suppose the lemma proved for $h-1$ rational functions and consider $\varphi_j(\mathbf{t})$ $(j \le h)$. By the inductive assumption there exist $r$ arithmetic progressions $P_\ell : at + b_\ell$ such that if $t_\ell^* \in P_\ell$ for all $\ell \le r$ then no $\varphi_j(t^*)$ is an integer of $K$ for $j < h$. Substituting progressions in question into $\varphi_h$ we get $\varphi_h(at_1+b_1, at_2+b_2,\ldots,at_r+b_r) = \psi_h(\mathbf{t})$, a rational function but not a polynomial. By the case $h = 1$ there exist progressions $ct + d_\ell$ such that if $t_\ell^* \equiv d_\ell \pmod{c}$ for all $\ell \ge r$ then $\psi_h(\mathbf{t}^*)$ is not an integer of $K$. The progressions $a(ct + d_\ell) + b_\ell$ have the property required in the lemma.

<u>Proof of Theorem 36</u>. Let $\langle t_1,\ldots,t_r\rangle = \mathbf{t}$,

$$F(x,\mathbf{t}) \underset{k[\mathbf{t}]}{\overset{\text{can}}{=}} a_0(\mathbf{t}) \prod_{\sigma=1}^{s_0} F_\sigma(x,\mathbf{t})^{e_\sigma},$$

where
$$|F_\sigma|_x \begin{cases} = 1 & \text{if} \quad \sigma \le s_1 \\ > 1 & \text{if} \quad \sigma > s_1 \end{cases}$$

and for $\sigma \le s_1$ denote by $x_\sigma(\mathbf{t})$ the rational function satisfying $F_\sigma(x_\sigma(\mathbf{t}),\mathbf{t}) = 0$. If for a certain $\sigma$ we have $x_\sigma(\mathbf{t}) \in K[\mathbf{t}]$ the theorem holds. If for all $\sigma \leq s_1$ we have $x_\sigma(\mathbf{t}) \notin K[\mathbf{t}]$ by Lemma we can choose $r$ arithmetic progressions $P_\ell : at+b_\ell$ such that for $\mathbf{t}^* \in P_1 \times P_2 \times \ldots \times P_r$ no $x_\sigma(\mathbf{t}^*)$ is an integer of $K$. By Theorem 34 there exist other arithmetic progressions $Q_\ell : ct+d_\ell$ such that if $\mathbf{t}^* \in Q_1 \times Q_2 \times \ldots \times Q_r$, then

$$\prod_{\sigma=s_1+1}^{s_0} F_\sigma(x,at_1^*+b_1,\ldots,at_r^*+b_r) \neq 0 \quad \text{for all} \quad x \in \mathbb{Q}.$$

The progressions $a(ct+d_\ell)+b_\ell$ $(\ell \leq r)$ have the property of contradicting the hypotheses of the theorem. Hence if the hypothesis holds $s_1 > 0$ and some $x_\sigma(\mathbf{t}) \in K[\mathbf{t}]$.

<u>Convention</u>. The fixed divisor of a polynomial over $\mathbb{Z}$ is the greatest common divisor of all its values at integer points.

<u>Corollary</u>. Let $F \in \mathbb{Z}[x,\mathbf{t}], \mathbf{t} = \langle t_1,\ldots,t_r\rangle$ and assume that the fixed divisor of $F$ equals $C(F)$, the content of $F$. If for every $r$ arithmetic progresions $P_1,\ldots,P_r$ there exists an integer vector $\mathbf{t}^* \in P_1 \times \ldots \times P_r$ and an $x^* \in \mathbb{Z}$ such that $F(x^*,\mathbf{t}^*) = 0$, then there exists a polynomial $X \in \mathbb{Z}[\mathbf{t}]$ such that $F(X(\mathbf{t}),\mathbf{t}) = 0$.

<u>Proof</u>. Let

$$F(x,\mathbf{t}) = F_0(x,\mathbf{t}) \prod_{\sigma=1}^{s} (x - X_\sigma(\mathbf{t})),$$

where the equation $F_0(x,\mathbf{t}) = 0$ has no solution in $\mathbb{Q}[\mathbf{t}]$ and $X_\sigma \in \mathbb{Q}[\mathbf{t}]$ $(1 \leq \sigma \leq s)$. Suppose that for all $\sigma \leq s$ we have $X_\sigma \notin \mathbb{Z}[\mathbf{t}]$. Then

$$C(x - X_\sigma(\mathbf{t})) = n_\sigma^{-1}, \quad n_\sigma \text{ an integer} > 1 \quad (1 \le \sigma \le s).$$

We have

$$F(x,\mathbf{t}) \prod_{\sigma=1}^{s} n_\sigma = F_0(x,\mathbf{t}) \prod_{\sigma=1}^{s} (n_\sigma x - n_\sigma X_\sigma(\mathbf{t})).$$

By the assumption about $F$ the content and the fixed divisor of the left hand side are equal and the same applies to the right hand side. Here the content is $C(F_0)$, hence the fixed divisor of $\prod_{\sigma=1}^{s} (n_\sigma x - n_\sigma X_\sigma(\mathbf{t}))$ is 1. Let $\prod_{\sigma=1}^{s} n_\sigma = a$. For every prime $p \mid a$ there exists an $x_p \in \mathbf{Z}$ and a $\mathbf{t}_p \in \mathbf{Z}^r$ such that

$$p \nmid \prod_{\sigma=1}^{s} (n_\sigma x_p - n_\sigma X_\sigma(\mathbf{t}_p)).$$

Choosing $x_0 \equiv x_p \pmod p$ and $\mathbf{b} \equiv \mathbf{t}_p \pmod p$ for all $p \mid a$ we infer that for all $\mathbf{t}^* \equiv \mathbf{b} \bmod a$

$$(a, \prod_{\sigma=1}^{s} (n_\sigma x_0 - n_\sigma X_\sigma(\mathbf{t}^*))) = 1,$$

thus none of the numbers $X_\sigma(\mathbf{t}^*)$ $(1 \le \sigma \le s)$ is an integer.

Now the equation $F_0(x, a\mathbf{t} + \mathbf{b}) = 0$ clearly has no solution in $Q[\mathbf{t}]$ and by Theorem 36 there exist integers $c$ and $d_\ell$ $(1 \le \ell \le r)$ such that if $t_\ell^* \equiv d_\ell \pmod c$ then the equation $F_0(x, a\mathbf{t}^* + \mathbf{b}) = 0$ has no solution in integers $x$.

The arithmetic progressions $a(ct + d_\ell) + b_\ell$ $(1 \le \ell \le r)$

have the property of contradicting the hypothesis of the corollary. Thus for at least one $\sigma \leq s$ we have $X_\sigma \in \mathbb{Z}[\mathbf{t}]$ and clearly $F(X_\sigma(\mathbf{t}),\mathbf{t}) = 0$.

Remark 1. A weaker form of Theorem 35 for $K = \mathbb{Q}$ in which the solvability of $F(x,t_1^*,\ldots,t_r^*) = 0$ is assumed for every $r$ integers $t_1^*,\ldots,t_r^*$ has been proved first by Kojima 1915. Under the same assumption he has also proved the special case $F(x,\mathbf{t}) = x^k - G(\mathbf{t})$ of Corollary.

Remark 2. Without the assumption about the fixed divisor of $F$ the conclusion of Corollary need not hold. This is shown by the examples $F_1 = 2x - t(t+1)$, $F_2 = (2x-t)(2x-t-1)$.

Theorems 34 and 36 suggest several problems. The simplest are following.

Problem 1. Let $F \in \mathbb{Q}[x,y,t_1,\ldots,t_r]$. If for every $r$ integers $t_1^*,\ldots,t_r^*$ the equation $F(x,y,t_1^*,\ldots,t_r^*) = 0$ has a solution $\langle x^*,y^*\rangle \in \mathbb{Q}^2$ do there exist rational functions $x,y \in \mathbb{Q}(t_1,\ldots,t_r)$ such that

$F(x(t_1,\ldots,t_r),y(t_1,\ldots,t_r),t_1,\ldots,t_r) = 0$ identically ?

Problem 2. The same with $\langle x^*,y^*\rangle \in \mathbb{Z}^2$, $x,y \in \mathbb{Q}[t_1,\ldots,t_r]$.

The answer to Problem 2 is no. For $r = 2$ it is enough to take $F(x,y,t_1,t_2) = t_1^2x + t_2^2y - t_1t_2$. (cf. Skolem 1940).

For $r = 1$ known counterexamples are more complicated. Take

$$F(x,y,t) = (x^2+1)(x^2+2)(x^2-2) - (2t-1)y.$$

For every $t^* \in \mathbb{Z}$ we have

$$2t^* - 1 = \pm \prod_{i=1}^{i_1} p_i^{\alpha_i} \prod_{i=1}^{i_2} q_i^{\beta_i} \prod_{i=1}^{i_3} r_i^{\gamma_i}$$

where $p_i, q_i, r_i$ are distinct primes satisfying the congruences $p_i \equiv 1 \bmod 4$, $q_i \equiv 3 \bmod 8$, $r_i \equiv 7 \bmod 8$.

We can find $a_i, b_i, c_i$ such that

$a_i^2 + 1 \equiv 0 \bmod p^{\alpha_i}$, $b_i^2 + 2 \equiv 0 \bmod q_i^{\beta_i}$, $c_i^2 - 2 \equiv 0 \bmod r_i^{\gamma_i}$.

Using Chinese remainder theorem we get an integer $x^*$ satisfying

$$x^* \equiv \begin{cases} a_i \bmod p_i & (1 \leq i \leq i_1), \\ b_i \bmod q_i & (1 \leq i \leq i_2), \\ c_i \bmod r_i & (1 \leq i \leq i_3). \end{cases}$$

For this $x^*$ the equation $F(x^*, y, t^*) = 0$ is solvable for $y$ in $\mathbf{Z}$. But there do not exist $x(t), y(t) \in \mathbb{Q}[t]$ such that

$$(x(t)^2 + 1)(x(t)^2 + 2)(x(t)^2 - 2) - (2t - 1)y(t) = 0.$$

Indeed the substitution $t = \frac{1}{2}$ gives $x(\frac{1}{2}) = \pm \zeta_4$ or $\pm \sqrt{-2}$ or $\pm \sqrt{2}$, which is impossible.

The answer to Problem 1 is probably also negative (see Lewis and Schinzel 1980), but the proof is lacking. One can at least show that an analog of Theorem 34 is false in general. Take $F(x,y,t) = y^2 - x^3 - 14t$. By a result of Mordell (see Mordell 1969, p.41) the congruence $y^2 - x^3 \equiv 14\,k \pmod{a}$ is solvable for all positive integers $a$ and $k$. Hence every arithmetic progression contains an integer $t^*$ such that $x^2 - y^3 = t^*$ is solvable in integers $x,y$. On the other hand no rational functions $x,y \in \mathbb{Q}(t)$ would satisfy $x^2 - y^3 = 14\,t$ identically. This can be seen on taking $t = s^6$ $(s = 1,2,\ldots)$ since the equation $y^2 - x^3 = 14s^6$ has no rational solution (ibid. p.250).

In section 25 we shall show an analog of Theorem 34 for quadratic equations with two unknowns. For this we need some preparation.

Theorem 37. (Davenport, Lewis and Schinzel 1966 for $K = \mathbb{Q}$, $r = 1$; Lewis and Schinzel 1980 for $K = \mathbb{Q}$, $r > 1$). *Let $K$ be an algebraic number field, $A, B \in K[t_1, \dots, t_r]$. Suppose that for every $r$ arithmetic progressions $P_1, \dots, P_r$ in $\mathbb{Z}$ there exist integers $t_1^*, \dots, t_r^*$ and integers $x^*, y^*, z^*$ of $K$ such that $t_l^* \in P_l$ $(1 \le l \le r)$, $(x^*, y^*, z^*) \ne (0,0,0)$ and*

$$(1)\quad A(t_1^*, \dots, t_r^*)x^{*2} + B(t_1^*, \dots, t_r^*)y^{*2} = z^{*2}.$$

*Then there exist polynomials $X, Y, Z \in K[t_1, \dots, t_r]$ not all zero such that*

$$(2)\quad AX^2 + BY^2 = Z^2.$$

Lemma 1. If $\boldsymbol{\tau} = \langle \tau_1, \dots, \tau_{r-1} \rangle \in \mathbb{Z}^{r-1}$ and $A(\boldsymbol{\tau}, t_r)$ is square-free then under the assumptions of the theorem there exists a polynomial $C(t_r) \in K[t_r]$ such that

$$(3)\quad B(\boldsymbol{\tau}, t_r) \equiv C(t_r)^2 \bmod A(\tau, t_r) \quad \text{and} \quad |C| < |A|_{t_r}.$$

Proof. We may assume that $A, B$ have integral coefficients in $K$ and replace $t_r$ by $t$. Let

$$A(\boldsymbol{\tau}, t) \underset{K}{\overset{\mathrm{can}}{=}} a_0 \prod_{\sigma=1}^{s} A_\sigma(t)$$

where $A_\sigma$ are monic for all $\sigma \le s$.

Take $\theta$ such that $A_1(\theta) = 0$. Choose in $K(\theta)$ a prime ideal of degree 1 dividing neither the discriminant of $K$ nor $a_0 D$, where $D$ is discriminant of $A(\tau, t)$. For a suitable

It follows that

$$0 = A_1(\theta) \equiv A_1(t_o) \bmod \mathfrak{Q}$$

and also $A_1(t_o) \equiv 0 \bmod \mathfrak{q}$ , where $\mathfrak{q} = N_{K/(\theta)/K}\mathfrak{Q}$ . We have for suitable $U, V \in K[t]$ with coefficients integral in $K$

$$U(t)A(\tau,t) + V(t)A'(\tau,t) = a_oD ,$$

where $A'$ is the derivative of $A$ with respect to $t_r$ . Since $\mathfrak{q} \nmid a_oD$ and $A_1(\tau,t_o) \equiv 0 \bmod \mathfrak{Q}$ , we have

$$A(\tau,t_o) \equiv 0 \bmod \mathfrak{q} , A'(\tau,t_o) \not\equiv 0 \bmod \mathfrak{q} .$$

Let $q = N_{K/\mathbb{Q}}\mathfrak{q}$ . Since $\mathfrak{q}$ does not divide the discriminant of $K$ , $\mathfrak{q}^2 \nmid q$ and we infer from the expansion

$$A(\tau,t_o+q) = A(\tau,t_o) + qA'(\tau,t_o)+ \cdots$$

that either $A(\tau,t_o) \not\equiv 0 \bmod \mathfrak{q}^2$ or $A(\tau_o,q) \not\equiv 0 \bmod \mathfrak{q}^2$ . Taking $\tau_r = t_o$ or $t_o + q$ respectively we get

(4) $$\tau_r \equiv \theta \bmod \mathfrak{Q} ,$$

(5) $$A(\tau,\tau_r) \equiv 0 \bmod \mathfrak{q} , A(\tau,\tau_r) \not\equiv 0 \bmod \mathfrak{q}^2 .$$

Let $e$ be the least positive exponent such that $\mathfrak{q}^e$ is principal in $K$ . By the assumption of the theorem applied to the arithmetic progressions $q^{2e}u + \tau_\ell (1 \le \ell \le r)$ there exist integers $t_\ell^*$ such that $t_\ell^* \equiv \tau_\ell \bmod \mathfrak{q}^{2e}$ $(1 \le \ell \le r)$ and suitable integers $x^*, y^*, z^*$ of $K$ not all $0$ such that (1) holds. If $x^*, y^*$ have a common ideal factor principal in $K$ say $(d)$ then we replace

$x^*, y^*, z^*$ in (1) by $x^*/d$, $y^*/d$, $z^*/d$. Thus we may assume that in (1) $(x^*, y^*)$ is not divisible by $q^e$.

From (1) it follows that

$$(6) \qquad A(\tau, \tau_r)x^{*2} + B(\tau, \tau_r)y^{*2} \equiv z^{*2} \bmod q^{2\ell} .$$

Let $x^*, y^*, z^*$ be divisible exactly by $q^\alpha, q^\beta, q^\gamma$ respectively, where $\alpha, \beta$ or $\gamma$ might be $\infty$, but $\min(\alpha, \beta) \le \min(\gamma, e-1)$.

If $\alpha < \beta$ then by (5) the left hand side of (6) is divisible exactly by $q^{2\alpha+1}$, while the right by $q^{2\gamma}$, which gives $2\alpha + 1 \ge 2e$, contrary to $\min\{\alpha, \beta\} \le e - 1$.

If $\alpha \ge \beta$, then also $\gamma \ge \beta$ and the congruence $y^*\omega \equiv z^* \bmod q^{2e}$ is solvable in $K$. We get from (6)

$$y^2(\omega^2 - B(\tau, \tau_r)) \equiv 0 \bmod q^{2\beta+1}$$

hence

$$\omega^2 - B(\tau, \tau_r) \equiv 0 \bmod q$$

and by (4)

$$\omega^2 - B(\tau, \theta) \equiv 0 \bmod \mathfrak{Q} .$$

Thus for almost all prime ideals $\mathfrak{Q}$ of degree 1 in $K(\theta)$ the above congruence is solvable for $\omega$ in $K(\theta)$. By the Lemma to Theorem 34 the binomial on the left hand side is reducible in $K(\theta)$, i.e.,.

$$B(\tau, \theta) = C_1(\theta)^2 , \text{ where } C_1 \in K[t] .$$

From the irreducibility of $A_1(t)$ it follows that

$$B(\tau,t) \equiv C_1(t)^2 \bmod A_1(t) .$$

By symmetry between $A_\sigma(t)$ we have similarly

$$B(\tau,t) \equiv C_\sigma(t)^2 \bmod A_\sigma(t) , C_\sigma \in K[t] (\sigma \leq s) .$$

By the Chinese remainder theorem for polynomials there exists a $C \equiv C_\sigma(t_r) \bmod A_\sigma(t_r)$ for all $\sigma \leq s$. It has required property (3) .

Lemma 2. Assume that $A$ is square-free with respect to $t_r$ and $|A|_{t_r} \geq |B|_{t_r}$. Under the assumptions of the theorem there exist polynomials $H(t_1,\cdots,t_{r-1}) \neq 0$ and $B_1(t_1,\cdots,t_r)$ over $K$ such that

$$H^2B \equiv B_1^2 \bmod A$$

in the ring $K[t_1,\cdots,t_r]$ and $|B_1|_{t_r} < |A|_{t_r}$.

Proof. For $r = 1$ the lemma is contained in Lemma 1. Thus we assume $r > 1$. Let $A$ viewed as a polynomial in $t_r$ have degree $a$, discriminant $D(\mathbf{t})$ and leading coefficient $a_0(\mathbf{t})$, where $\mathbf{t} = \langle t_1,\ldots,t_{r-1}\rangle$. By the hypothesis of the lemma $D(\mathbf{t}) \neq 0$.

Take in Theorem 16 $M = A$, $F = v^2 - B(\mathbf{t},t_r)$ and let $\Phi(v,\mathbf{t},t_r)$ be a non-zero polynomial the existence of which is asserted in that theorem. Further let

$$(7) \qquad \Phi(v,\mathbf{t},t_r) = \Phi_0(\mathbf{t},t_r) \prod_{\rho=1}^{q} \Phi_\rho(v,\mathbf{t},t_r),$$

where $\Phi_0 \in K[\mathbf{t},t_r]$, $\Phi_\rho \in K[v,\mathbf{t},t_r]$ $(1 \leq \rho \leq q)$ and for $\rho > 0$ the polynomials $\Phi_\rho$ are irreducible over $K$, have leading coefficient $\Psi_\rho(\mathbf{t},t_r)$ and positive degree with respect to $v$. We order the $\Phi_\rho$ so that $\Phi_\rho$ is of degree 1 in $v$ for $\rho \leq p$ and of degree at least 2 for $\rho > p$ and denote leading coefficient of $\Psi_\rho(\mathbf{t},t_r)$ with respect to $t_r$ by $\Psi_\rho(\mathbf{t})$. If for all $\rho \leq p$ we have

$$G_\rho(\mathbf{t},t_r) = B(\mathbf{t},t_r)\Psi_\rho(\mathbf{t},t_r)^2 - \Phi_\rho(0,\mathbf{t},t_r)^2 \not\equiv 0 \mod A(\mathbf{t},t_r)$$

then let the leading coefficient of the remainder on division of $G_\rho$ by $A(\mathbf{t},t_r)$ in the ring $\mathbb{K}(\mathbf{t})[t_r]$ be $g_\rho(\mathbf{t})a_0(\mathbf{t})^{-m_\rho}$, where $g_\rho \in K[\mathbf{t}]$.
By Theorem 33 there exists an integral vector $\mathbf{t}^* \in \mathbb{Z}^{r-1}$ such that all polynomials $\Phi_\rho(v,\mathbf{t}^*,t_r)$ are irreducible and

$$(8) \qquad a_0(\mathbf{t}^*)D(\mathbf{t}^*)\prod_{\rho=0}^{q}\Psi_\rho(\mathbf{t}\ )\prod_{\rho=1}^{p}g_\rho(\mathbf{t}^*) \neq 0.$$

Clearly $A(\mathbf{t}^*,t_r)$ is squarefree. It follows from Lemma 1 that there exists a polynomial $C \in K[t_r]$ satisfying (3). Then Theorem 16 gives

$$\Phi(C(t_r),\mathbf{t}^*,t_r) = 0.$$

It follows from (7) and the irreducibility of $\Phi_\rho(v,\mathbf{t}^*,t_r)$ for all $\rho > 0$ that for some $\rho \leq p$ we have

$$\Phi_\rho(C(t_r),\mathbf{t}^*,t_r) = \Psi_\rho(\mathbf{t}^*,t_r)C(t_r) + \Phi_\rho(0,\mathbf{t}^*,t_r) = 0.$$

Hence by (3)

$$\Psi_\rho(\mathbf{t}^*,t_r)^2B(\mathbf{t}^*,t_r) - \Phi_\rho(0,\mathbf{t}^*,t_r)^2 \equiv 0 \mod A(\mathbf{t}^*,t_r)$$

and so $g_\rho(\mathbf{t}^*) = 0$ contrary to (8). The obtained contradiction shows that for a certain $\rho \leq p$

$$\Psi_\rho(\mathbf{t},t_r)^2 B(\mathbf{t},t_r) - \Phi_\rho(0,\mathbf{t},t_r)^2 \equiv 0 \mod A(\mathbf{t},t_r).$$

By the irreducibility of $\Phi_\rho(v,\mathbf{t},t_r)$ we have $(\Psi_\rho(\mathbf{t},t_r),\Phi_\rho(0,\mathbf{t},t_r)) = 1$ and hence there exists a $\beta \in K(\mathbf{t})[t_r]$ such that

$$B(\mathbf{t},t_r) \equiv \beta^2(\mathbf{t},t_r) \mod A(\mathbf{t},t_r)$$

in the ring $K(\mathbf{t})[t_r]$ and moreover $|\beta|_{t_r} < a$ (including the possibility $\beta = 0$). Let

$$\beta^2(\mathbf{t},t_r) - B(\mathbf{t},t_r) = H^{-2}(\mathbf{t})A(\mathbf{t},t_r)A_1(\mathbf{t},t_r),$$

where $H \in K[\mathbf{t}]$ and $A_1(\mathbf{t},t_r) \in K[\mathbf{t},t_r]$.

We have $H \neq 0$, $B_1 = H\beta \in K[\mathbf{t},t_r]$ and $H^2B \equiv B_1^2 \mod A$, as asserted in the lemma. Observe further that

$$|B_1|_{t_r} = |\beta|_{t_r} < a.$$

Proof of Theorem 37. We proceed by induction on $r$. For $r = 0$ the theorem in trivial. Suppose that $r \geq 1$ and that the theorem is true for fewer than $r$ parameters. If $A$ or $B$ is identically $0$ we take $Z = 0$ and $X = 0$, $Y = 1$ or $X = 1$, $Y = 0$ respectively.

We now proceed by induction on the degree of $AB$ with respect to $t_r$, denoted for simplicity by $|AB|_r$. Suppose the result holds for all $A,B$ satisfying $|A|_r + |B|_r < n$, where $n$ is some positive integer; we have to prove the result for polynomials $A,B$, when $|A|_r + |B|_r = n$. We can suppose without loss of generality that $a = |A|_r \geq |B|_r$ and so in particular $a > 0$. Suppose first that $A$ is not square-free as a polynomial in $t_r$, say

$$A = C_o^2 A_o, \quad A_o C_o \in K[t_1,\ldots,t_r], \quad |C_o|_r \geq 1.$$

The hypothesis of the theorem regarding $A,B$ insures that the hypothesis also holds for polynomials $A_o,B$. Indeed, for every $r$ arithmetic progressions $P_1,\ldots,P_r$ there exist, e.g. by Theorem 34, with $F = C_o$, arithmetic progressions $P'_1,\ldots,P'_r$ such that $P'_\ell \subset P_\ell$ $(\ell \leq r)$ and if $t^*_\ell \in P'_\ell$ for all $\ell \leq r$ then $C_o(t^*_1,\ldots,t^*_r) \neq 0$. The hypothesis for $A,B$ implies there are integers $t^*_\ell$ $(\ell \leq r)$ and elements $x^*,y^*,z^*$ of $K$ satisfying $t^*_\ell \in P_\ell$ for all $\ell \leq r$, $\langle x^*,y^*,z^*\rangle \neq \langle 0,0,0\rangle$ and (1). But then $t^*_\ell \in P_\ell$ $(\ell \leq r)$ and

$$A_o(t^*_1,\ldots,t^*_r)x^2 + B(t^*_1,\ldots,t^*_r)y^2 = z^2$$

has $x^* C_o(t^*_1,\ldots,t^*_r), y^*, z^*$ as a nontrivial solution in $K$. Since $|A_o|_r + |B|_r < |A|_r + |B|_r = n$ the inductive assumption implies the existence of polynomials $X_o,Y_o,Z_o \in K[t_1,\ldots,t_r]$ not all identically $0$ such that

$$A_o X_o^2 + BY_o^2 = Z_o^2.$$

On taking

$$X = X_o, \quad Y = Y_oC_o, \quad Z = Z_oC_o$$

we obtain an identical solution of (2).

Thus we can suppose that $A$ is square-free as a polynomial in $t_r$. By Lemma 2 there exist polynomials $H, A_1, B_1$ such that $H \in K[\mathbf{t}], A_1, B_1 \in K[\mathbf{t}, t], H \neq 0$ and

$$(16) \qquad B_1^2 = H^2B + AA_1, \quad |B_1|_r < |A|,$$

where $\mathbf{t} = \langle t_1, \ldots, t_{r-1}\rangle$.

If $A_1(\mathbf{t}, t_r) = 0$, we can satisfy (2) by taking

$$X(\mathbf{t}, t_r) = 0, \quad Y(\mathbf{t}, t_r) = H(\mathbf{t}), \quad Z(\mathbf{t}, t_r) = B_1(\mathbf{t}).$$

If $A_1 \neq 0$ we have by (16) that $|A_1|_r < a$.

We now prove the hypothesis of the theorem are satisfied for the polynomials $A_1, B$. By Theorem 34 with $F = AA_1$ for every $r$ arithmetic progressions $P_1, \ldots, P_r$ there exist arithmetic progressions $P'_1, \ldots, P'_r$ such that $P'_\ell \subset P_\ell$ and if

$$t^*_\ell \in P'_\ell \quad (\ell \leq r) \quad \text{then}$$

$$(17) \qquad A(\mathbf{t}^*, t^*_r)A_1(\mathbf{t}^*, t^*_r) \neq 0.$$

On the other hand by the hypothesis of the theorem for $A, B$ there exist integers $t^*_\ell \in P'_\ell$ $(\ell \leq r)$ and integers $x^*, y^*, z^*$ of $K$ satisfying (1). Taking

$$x_1^* = A(\mathbf{t}^*, t_r^*)x^*, \quad y_1^* = H(\mathbf{t}^*)z^* = B_1(\mathbf{t}^*, t_r^*)y^*,$$

$$z_1^* = H(\mathbf{t}^*)B(\mathbf{t}^*, t_r^*)y^* - B_1(\mathbf{t}^*, t_r^*)z^*$$

we obtain $t_\ell^* \in P_\ell$ $(1 \leq \ell \leq r)$ and

$$A_1(\mathbf{t}^*, t_r^*)x_1^{*2} + B(\mathbf{t}^*, t_r^*)y_1^{*2} - z_1^{*2} =$$

$$= (B_1(\mathbf{t}^*, t_r^*)^2 - H(\mathbf{t}^*)^2 B(\mathbf{t}^*, t_r^*))(A(\mathbf{t}^*, t_r^*)x^{*2} + B(\mathbf{t}^*, t_r^*)y^{*2} - z^{*2}) = 0.$$

Also $x_1^*, y_1^*, z_1^*$ are not all $0$, since (17) holds. Without loss of generality we may assume that $x^*, y^*, z^*$ are integers of $K$.

The inductive hypothesis applies to the polynomials $A_1, B$ since $|A_1|_r + |B|_r < |A|_r + |B|_r = n$. Hence there exist polynomials $X_1, Y_1, Z_1 \in K[\mathbf{t}, t_r]$ not all zero such that

$$A_1X_1^2 + B_1Y_1^2 = Z_1^2.$$

Putting

$$X = A_1X_1, \quad Y = B_1Y_1 + HZ_1, \quad Z = HBY + B_1Z$$

we obtain (2). Further $X, Y, Z$ do not all vanish identically since $AA_1 \neq 0$. The inductive proof is complete.

Theorem 38 (Davenport, Lewis and Schinzel 1966 for $K = \mathbb{Q}$, $r = 1$, Lewis and Schinzel 1980 for $K = \mathbb{Q}$, $r > 1$). *Let $K$ be an algebraic number field, $F \in K[x,y,t_1,\dots,t_r]$ be of degree at most 2 in $x$ and $y$. If for every $r$ arithmetic progressions $P_1,\dots,P_r$ in $\mathbb{Z}$ there exist integers $t_1^*,\dots,t_r^*$ and elements $x^*,y^*$ of $K$ such that $t_l^* \in P_l$ $(1 \le l \le r)$ and*

$$(1) \qquad F(x^*,y^*,t_1^*,\dots,t_r^*) = 0$$

*then there exist rational functions $x,y \in K(t_1,\dots,t_r)$ such that*

$$(2) \qquad f(x(t_1,\dots,t_r),y(t_1,\dots,t_r),t_1,\dots,t_r) = 0.$$

Lemma. The theorem is true for $F(x,y,\mathbf{t}) = A(\mathbf{t})x^2+B(\mathbf{t})y^2+C(\mathbf{t})$, where $\mathbf{t} = \langle t_1,\dots,t_r \rangle$.

Proof. If $C(\mathbf{t}) = 0$, then $x(\mathbf{t}) = y(\mathbf{t}) = 0$ is an identical solution.

If $C(\mathbf{t}) \neq 0$ then for every $r$ arithmetic progressions $P_1,\dots,P_r$ there exist sub-progressions $P_1',\dots,P_r'$ such that if $\mathbf{t}^* \in P_1' \times \dots \times P_r'$ then $C(\mathbf{t}^*) \neq 0$. Taking

$$A_0(\mathbf{t}) = -A(\mathbf{t})C(\mathbf{t}), \quad B_0(\mathbf{t}) = -B(\mathbf{t})C(\mathbf{t})$$

and applying the assumption of the theorem to the arithmetic progressions $P_1',\ldots,P_r'$ we infer the existence of a $\mathbf{t}^* \in P_1' \times \ldots \times P_r'$, $x^*, y^* \in K$ such that

$$(3) \qquad A_o(\mathbf{t}^*)x^{*2} + B_o(\mathbf{t}^*)y^{*2} = C(\mathbf{t}^*)^2.$$

We have $\mathbf{t}^* \in P_1 \times \ldots \times P_r$, $C(\mathbf{t}^*) \neq 0$ and by Theorem 37 there exist polynomials $X(\mathbf{t}), Y(\mathbf{t}), Z(\mathbf{t})$ not all $0$ such that

$$(4) \qquad A_oX^2 + B_oY^2 = Z^2.$$

Now if $Z(\mathbf{t}) \neq 0$ we take

$$x(t) = \frac{X(\mathbf{t})}{Z(\mathbf{t})}, \quad y(t) = \frac{Y(\mathbf{t})}{Z(\mathbf{t})}.$$

If $Z(\mathbf{t}) = 0$ but $Y(\mathbf{t}) \neq 0$ we have

$$A_oX^2 + B_oY^2 = 0$$

so

$$B_o = -A_o\left(\frac{X}{Y}\right)^2.$$

If $A_o = 0$ then $B_o = 0$, hence by (3) $C(\mathbf{t}^*) = 0$ contrary to the choice of $\mathbf{t}^*$. Thus $A_o \neq 0$. Choose now the rational functions $x, y$ so that

$$x - \frac{X}{Y}y = A_o^{-1}C, \quad x + \frac{X}{Y}y = C.$$

Then

$$Ax^2 + By^2 = C^{-1}(A_o x^2 + B_o y^2) = C^{-1}A_o(x - \frac{X}{Y}y)(x + \frac{X}{Y}y) = C.$$

If $Z(t) = Y(t) = 0$ then by (4) $A_o = 0$. The equation (3) gives

$$B_o(\mathbf{t}^*)y^{*2} = C(t^*)^2 \neq 0$$

hence $B_o(\mathbf{t}^*) = (\frac{C(\mathbf{t}^*)}{y^*})^2$.

By Theorem 34 $B_o = C_o^2$ with $C_o \in K(\mathbf{t})$. We take

$$x(\mathbf{t}) = 0, \quad y(\mathbf{t}) = \frac{C(\mathbf{t})}{C_o(\mathbf{t})} .$$

Proof of Theorem 38. If $F$ does not depend on $x$, the theorem holds by virtue of Theorem 34. If $F$ is of degree 1 in $x$, the theorem holds trivially. If $F$ is of degree 2 in $x$, we can write either

$$(5) \qquad F(x,y,\mathbf{t}) = a(t)(x+\alpha(\mathbf{t})y+\beta(\mathbf{t}))^2+b(\mathbf{t})(y+\gamma(\mathbf{t}))^2+c(\mathbf{t})$$

or

$$(6) \qquad F(x,y,\mathbf{t}) = a(\mathbf{t})(x+\alpha(\mathbf{t})y+\beta(t))^2+b_1(\mathbf{t})y+c(\mathbf{t}),$$

where $a,b,b_1,c,\alpha,\beta,\gamma \in K(\mathbf{t})$ and $ab_1 \neq 0$.

In case (6) we can satisfy (2) by taking

$$y(\mathbf{t}) = -\frac{c(\mathbf{t})}{b_1(\mathbf{t})}, \quad x(\mathbf{t}) = -\alpha(\mathbf{t})y(\mathbf{t})-\beta(\mathbf{t}).$$

In case (5) we write

$$a(t) = \frac{A(t)}{D(t)}, \quad b(t) = \frac{B(t)}{D(t)}, \quad c(t) = \frac{C(t)}{D(t)},$$

where $A,B,C,D \in K[t]$, $AD \neq 0$.

The equality (1) implies

$$A(t^*)(x^*+\alpha(t^*)y^*+\beta(t^*))^2+B(t^*)(y^*+\gamma(t^*))^2+C(t^*) = 0,$$

thus by the lemma there exist rational functions $x_o, y_o \in K(t)$, such that

$$Ax_o^2 + By_o^2 + C = 0.$$

We now satisfy (2) by taking

$$y(t) = y_o(t)-\gamma(t), x(t) = x_o(t)-\alpha(t)y(t)-\beta(t).$$

The proof is complete.

The following example shows that Theorem 38 does not extend to quadratic equations with three unknowns. By Gauss's theorem (see Mordell 1969, p. 175) the equation

$$x^2 + y^2 + z^2 = 28t^{*^2} + 1$$

is solvable in integers $x,y,z$ for every integer $t^*$. On the other hand if rational functions $x(t), y(t), z(t) \in \mathbb{Q}(t)$ satisfied the equation

$$x(t)^2 + y(t)^2 + z(t)^2 = 28t^2 + 1$$

we would have

$$a^2 + b^2 + c^2 = 28, \quad a,b,c \in \mathbb{Q}$$

where $a,b,c$ are the leading coefficients of the Laurent expansions of $x(t),y(t),z(t)$ in the neighbourhood of $\infty$. This, however is impossible by the same Gauss's theorem.

There exist also equations $F(x,y,t) = 0$ quadratic with respect to $x$ and $y$ and solvable in integers $x,y$ for all integers $t$ but unsolvable in polynomials $x,y \in \mathbb{Q}[t]$. An example is

$$x^2 - (4t^2 + 1)^3 y^2 = -1.$$

For the proof (not very difficult) see the forthcoming paper of the author (Schinzel 1981). In the same paper it is also proved that if $L \in \mathbb{Q}[x,t]$ is of degree at most four in $x$, $M \in \mathbb{Q}[t]$ is arbitrary and every arithmetic progression contains an integer $t^*$ such that $L(x,t^*) = M(t^*)y$ is solvable in integers $x,y$ then there exist polynomials $X,Y \in \mathbb{Q}[t]$ such that $L(X(t),t) = M(t)Y(t)$. The result does not extend to polynomials $L$ of degree five in $x$.

Taking in Theorem 38 $F(x,y,\mathbf{t}) = A(\mathbf{t})(x^2+y^2) + C(\mathbf{t})$ we infer that a rational function over an algebraic number field $K$ whose values are norms from $K(\zeta_4)$ to $K$ is itself a norm from $K(\zeta_4,\mathbf{t})$ to $K(\mathbf{t})$. In the present section we shall prove as Theorem 40 an extension of this result. As a preparation we need

<u>Theorem 39.</u> (Kronecker - A. Kneser 1887). *Let $A,B$ be polynomials irreducible and separable over a field $k$ and $A(\alpha) = B(\beta) = 0$. If*

$$A(x) \underset{k(\beta)}{\overset{can}{=}} \prod_{\rho=1}^{r} A_\rho(x) \,,$$

(1)

$$B(x) \underset{k(\alpha)}{\overset{can}{=}} \prod_{\sigma=1}^{s} B_\sigma(x)$$

*then $r = s$, and after a suitable renumbering*

$$\frac{|A_\rho|}{|A|} = \frac{|B_\rho|}{|B|} \quad (1 \leq \rho \leq r) \,.$$

Proof. (Bauer 1929). Write more explicitly
$A_\rho(X) = A_\rho(\beta,x)$, $B_\sigma(x) = B_\sigma(\alpha,x)$, where
$|A_\rho|_\beta < |B|$, $|B_\sigma|_\alpha < |A|$. By Lemma 2 to Theorem 33

$$N_{k(\beta)/k} A_\rho(\beta,x) = a_\rho A(x)^{m_\rho} \text{ for some } a_\rho \in k \,.$$

Comparing degrees of both sides we get

(2) $$|B| \, |A_\rho|_x = m_\rho |A| \,.$$

Comparing the multiplicity of $\alpha$ as zero of both sides we infer that

$$(3) \qquad m_\rho = \# \{\nu \leq |B| \ , A_\rho(\beta_\nu, \alpha) = 0\} \ ,$$

$\beta_\nu$ being the conjugates of $\beta$ over $k$. Form new polynomials

$$C_\rho(x) = (B(x) \ , A_\rho(x,\alpha)) \ (1 \leq \rho \leq r) \ .$$

Then $C_\rho(x) \neq$ const since there exists a $\nu \leq |B|$ such that $A_\rho(\beta_\nu, \alpha) = 0$. On the other hand, for $\rho \neq \rho'$ we have $(C_\rho, C_{\rho'}) = 1$, otherwise for a $\nu \leq |B|$ $C_\rho(\beta_\nu) = C_{\rho'}(\beta_\nu) = 0$, so the two factors $A_\rho(\beta_\nu, x)$, $A_{\rho'}(\beta_\nu, x)$ would have common root $\alpha$, which is impossible by (1). We have found $r$ pairwise relatively prime divisors of $B$ over $k(\alpha)$, hence $r \leq s$. By symmetry $s \leq r$ so $r = s$. Moreover $C_\rho$ are up to a constant factor the same as $B_\sigma$ in a possibly different order. After renumbering

$$C_\rho(x) = (B(x) \ , A_\rho(x,\alpha)) = B_\rho(\alpha, x) \ .$$

By symmetrical argument, after another possibly different renumbering

$$(A(x) \ , B_\rho(x,p)) = A_\rho(\beta, x) \ .$$

Now $\alpha$ is a root of $A_\rho(\beta_\nu, x)$ if and only if $B_\rho(\alpha, \beta_\nu) = 0$. Hence by (3)

$$m_\rho = \# \{\nu \leq |B| : B_\rho(\alpha, \beta_\nu) = 0\} = |B_\rho|_x \ .$$

This together with (2) proves the theorem.

Now we can prove

**Theorem 40.** (Schinzel 1966 for $k = \mathbb{Q}$, $r = 1$, Schinzel 1973a for $k = \mathbb{Q}$, $r > 1$). *Let $K$ be an extension of prime degree of an algebraic number field $k$, and let $f \in k(t_1, \cdots, t_r)$. If for every $r$ arithmetic progressions $P_1, \cdots, P_r$ in $\mathbb{Z}$ there exists $t_1^*, \cdots, t_r^*$ and $\omega^*$ such that*

$$t_\ell^* \in P_\ell (1 \leq \ell \leq r), \quad \omega^* \in K, \tag{4}$$

$$f(t_1^*, \cdots, t_r^*) = N_{K/k}(\omega^*),$$

*then there exists an $\omega(t_1, \cdots, t_r) \in K(t_1, \cdots, t_r)$ such that*

$$f = N_{K/k}\, \omega(t_1, \cdots, t_r). \tag{5}$$

**Lemma 1.** Let $K$ be a finite extension of an algebraic number field $k$, $f \in k(t_1, \cdots, t_r)$, $f \neq 0$ and $F$ be an irreducible polynomial with integral coefficients in $k$ such that

$$\operatorname{ord}_F f = e \neq 0, \tag{6}$$

$$F \underset{K}{\overset{\text{can}}{=}} c \prod_{\nu=1}^{n} \Phi_\nu(t_1, \cdots, t_r). \tag{7}$$

If for every $r$ arithmetic progressions $P_1, \cdots, P_r$ in $\mathbb{Z}$ there exist $t_1^*, \cdots, t_r^*$ and $\omega^*$ such that (4) holds then there exists an integral vector $\boldsymbol{\tau} \in \mathbb{Z}^{r-1}$ such that

(8) $\Phi_\nu(\boldsymbol{\tau}, t)$ are irreducible over $K (1 \leq \nu \leq n)$,

(9) $F(\boldsymbol{\tau}, t)$ is irreducible over $k$,

(10) for almost all prime ideals $\mathfrak{q}$ of $k$ such that $F(\boldsymbol{\tau}, t) \equiv 0 \bmod \mathfrak{q}$ is solvable for $t$ in $\mathbb{Z}$ there

exists an $\omega \in K$ such that

$$\text{(11)} \qquad \operatorname{ord}_{\mathfrak{q}} N_{K/k}(\omega) = e .$$

Proof. Let $\langle t_1, \cdots, t_r \rangle = \mathbf{t}$

$$\text{(12)} \qquad f(\mathbf{t}) \underset{k}{\overset{\text{can}}{=}} a \prod_{\sigma=1}^{s} F_\sigma(\mathbf{t})^{e_\sigma}, \; e_\sigma \in \mathbb{Z} .$$

We may assume without loss of generality that $F = F_1$ and that all $F_\sigma$ have coefficients integral in $k$. Let $R(t_1, \cdots, t_{r-1})$ be the resultant of $F$ and $\frac{\partial F}{\partial t_r} \cdot \prod_{\sigma=2}^{s} F_\sigma(\mathbf{t})$ with respect to $t_r$. We have $R(t_1, \cdots, t_{r-1}) \neq 0$ and there exist polynomials $U, V \in k[t_1, \cdots, t_r]$ with coefficients integral in $k$ such that

$$\text{(13)} \qquad UF + V \frac{\partial F}{\partial t_r} \prod_{\sigma=2}^{s} F_\sigma = R .$$

By Theorem 33 there exists an integral vector $\boldsymbol{\tau} \in \mathbb{Z}^{r-1}$ such that (8) holds and

$$\text{(14)} \qquad R(\boldsymbol{\tau}) \neq 0 .$$

By (7) of Lemma 2 to Theorem 33 $N_{K/k}\Phi_1(\mathbf{t}) = cF(\mathbf{t})^m$ hence by substitution

$$N_{K/k}\Phi_1(\boldsymbol{\tau}, t) = c\,F(\boldsymbol{\tau}, t)^m .$$

On the other hand by the same lemma I $\text{const} \cdot N_{K/k}\Phi_1(\boldsymbol{\tau}, t)$ is a power of a polynomial $\Psi$ irreducible over $k$. Hence $\text{const} \cdot F(\boldsymbol{\tau}, t)$ is a power of $\Psi$. By (13) and (14) $F(\boldsymbol{\tau}, t)$ has no multiple factors, hence (9) holds. In order to prove (10) take $\mathfrak{q}$ to be any prime ideal of $k$

such that

(15) $$\mathfrak{q} \nmid DR(\boldsymbol{\tau}) \, , \ \mathrm{ord}_{\mathfrak{q}} a = 0 \, ,$$

where $D$ is the discriminant of $k$ and suppose that $F(\boldsymbol{\tau}, t_o) \equiv 0 \bmod \mathfrak{q}$ . By (13) we have

$$U(\boldsymbol{\tau}, t_o) F(\boldsymbol{\tau}, t_o) + V(\boldsymbol{\tau}, t_o) F'(\boldsymbol{\tau}, t_o) \prod_{\sigma=2}^{s} F_\sigma(\boldsymbol{\tau}, t_o) = R(\boldsymbol{\tau})$$

and since $U(\boldsymbol{\tau}, t_o) , V(\boldsymbol{\tau}, t_o)$ are integers of $k$

$$F'(\boldsymbol{\tau}, t_o) \prod_{\sigma=2}^{s} F_\sigma(\boldsymbol{\tau}, t_o) \not\equiv 0 \bmod \mathfrak{q} \ .$$

Let $q$ be the rational prime divisible by $\mathfrak{q}$ . By (15) $\mathfrak{q}^2 \nmid q$ .

It follows from the expansion

$$F(\boldsymbol{\tau}, t_o + q) = F(\boldsymbol{\tau}, t_o) + q \ F'(\boldsymbol{\tau}, t_o) + \cdots$$

that either $\mathfrak{q}^2 \nmid F(\boldsymbol{\tau}, t_o)$ or $\mathfrak{q}^2 \nmid F(\boldsymbol{\tau}, t_o + q)$ . Choosing $\boldsymbol{\tau}_r = t_o$ or $t_o + q$ accordingly, we get

$$\mathrm{ord}_{\mathfrak{q}} F(\boldsymbol{\tau}, \tau_r) = 1 \, , \ \mathrm{ord}_{\mathfrak{q}} F_\sigma(\boldsymbol{\tau}, \tau_r) = 0 (2 \le \sigma < s) \ .$$

Consider the arithmetic progressions $P_\ell : q^2 u + \tau_\ell (1 \le \ell \le r)$ . If $\mathbf{t}^* \in P_1 \times \cdots \times P_r$ we have

$$\mathrm{ord}_{\mathfrak{q}} F(\mathbf{t}^*) = 1 \ , \ \mathrm{ord}_{\mathfrak{q}} F_\sigma(\mathbf{t}^*) = 0 \ (2 \le \sigma \le s)$$

and by (12) and (15)

$$\mathrm{ord}_{\mathfrak{q}} f(\mathbf{t}^*) = e \ .$$

The equality (11) follows now from (4) for $\omega = \omega^*$ and (10) holds for this $\omega$ .

Lemma 2. Let $A(x)$ be a polynomial of prime degree irreducible over an algebraic number field $L$ . Then there exist infinitely many prime ideals $\mathfrak{Q}$ of degree 1 in $L$ such that $A$ is irreducible $\bmod \mathfrak{Q}$ .

Proof. If $A$ is irreducible of prime degree $p$ , the Galois group $\mathcal{G}$ of $A$ over $L$ is transitive on the zeros of $A$ , so $|\mathcal{G}| \equiv 0 \bmod p$ and $\mathcal{G}$ contains a p-cycle. By the Frobenius density theorem there exist infinitely many prime ideals $\mathfrak{Q}$ of degree 1 in $L$ such that $A$ is irreducible $\bmod \mathfrak{Q}$ .

Lemma 3. Let $K$ be of prime degree $p$ over an algebraic number field $k$ , $B \in k[t]$ be irreducible over $k$ with integral coefficients in $k$ , and $e$ an integer. If for almost all prime ideals $\mathfrak{q}$ of $k$ such that $B(t) \equiv 0 \bmod \mathfrak{q}$ is solvable in $\mathbb{Z}$ there exists an $\omega \in K$ satisfying

$$\operatorname{ord}_{\mathfrak{q}} N_{K/k}(\omega) = e$$

then $B(t)^e = b\, N_{K/k}(\omega(t))$ for some $\omega \in K(t)$ , $b \in k$ .

Proof. If $e \equiv 0 \bmod p$ take $\omega(t) = B(t)^{e/p}$ . If $e \not\equiv 0 \pmod p$ , let $b_0$ be the leading coefficient of $B$ . Take $\mathfrak{Q}$ a prime ideal of degree 1 in $L = k(\beta)$ , where $B(\beta) = 0$ . If $\mathfrak{Q} \nmid b_0$ we have $\beta \equiv t_0 \bmod \mathfrak{Q}$, $t_0 \in \mathbb{Z}$ , so $B(t_0) \equiv 0 (\bmod \mathfrak{Q})$ , i.e., $B(t_0) \equiv 0 \bmod \mathfrak{q}$ , where $\mathfrak{q}$ is the prime ideal of $k$ divisible by $\mathfrak{Q}$ .

By the assumption for almost all $\mathfrak{q}$ arising in this way there exists an $\omega \in K$ such that $\operatorname{ord}_{\mathfrak{q}} N_{K/k}(\omega) = e$ .

Since $(e,[K:k]) = 1$ it follows that $\mathfrak{q}$ factors in $K$. Let $K = k(\alpha)$, where $A(\alpha) = 0$ and $A$ is irreducible over $k$. By Dedekind's theorem (see Mann 1955, Theorem 16.2) apart from finitely many exceptional $\mathfrak{q}$'s $A(x)$ factors mod $\mathfrak{q}$ and hence also mod $\mathfrak{Q}$. Therefore by Lemma 2 $A(x)$ factors over $L$. By Theorem 39, $B(x)$ factors over $K$.

Let

$$A(x) \underset{L}{\overset{\text{can}}{=}} A_1, \cdots, A_r ,$$

$$B(x) \underset{K}{\overset{\text{can}}{=}} B_1, \cdots, B_r .$$

Then after a suitable renumbering we have

$$\frac{|A_\rho|}{|A|} = \frac{|B_\rho|}{|B|} \quad (1 \leq \rho \leq r)$$

and $N_{K/k} B_\rho = b_\rho B^{|A_\rho|}$, $|A_\rho| < p$.

We find $u$, $v \in \mathbf{Z}$ such that

$$pu + |A_1| v = e .$$

Then $N_{K/k}(B^u B_1^v) = B(x)^{up} b_1^v B(x)^{v|A_1|} = b_1^v B(x)^e$.

Proof of Theorem 40. If $f(\mathbf{t}) = 0$ the theorem is trivially true. If $f(t) \neq 0$ let the canonical factorization of $f$ in $k$ be given by (12), i.e., $f(\mathbf{t}) \underset{k}{\overset{\text{can}}{=}} a \prod_{\sigma=1}^{s} F_\sigma(t)^{e_\sigma}$ and let for a fixed $\sigma \leq s$

$$F_\sigma(\mathbf{t}) \underset{K}{\overset{\text{can}}{=}} c \prod_{\nu=1}^{n} \Phi_\nu(\mathbf{t}) , \tag{16}$$

$$(17) \qquad N_{K/k}\, \Phi_\nu(\mathbf{t}) = c_\nu F_\sigma(\mathbf{t})^{m_\nu} .$$

Since $\sigma$ is fixed we have omitted it as a subscript of $\Phi_\nu$ and $m_\nu$. By Lemma 1 there exists an integral vector $\boldsymbol{\tau} \in \mathbb{Z}^{r-1}$ such that (8), (9) and (10) hold for $F = F_\sigma$. Taking in Lemma 3 $B(t) = F_\sigma(\boldsymbol{\tau}, t)$ we infer that

$$(18) \qquad B(t)^{e_\sigma} = b\, N_{K/k}(\omega(t)) \quad \text{for some } \omega \in K(t),\ b \in k .$$

On the other hand by (16) $B(t) \underset{K}{\overset{can}{=}} c \prod_{\nu=1}^{n} \Phi_\nu(\boldsymbol{\tau}, t)$. If

$$\omega(t) \underset{K}{\overset{can}{=}} \beta \prod_{\nu=1}^{n} \Phi_\nu(\boldsymbol{\tau}, t)^{\eta_\nu} \prod_{(\chi,B)=1} \chi(t)^{e(\chi)}$$

we get from (17) and (18)

$$B(t)^{e_\sigma} = b\, N_{K/k}\, \beta \prod_{\nu=1}^{n} c_\nu^{\eta_\nu} F_\sigma^{m_\nu \eta_\nu} \prod_{(\chi,B)=1} N_{K/k} \chi(t)^{e(\chi)} .$$

Thus

$$e_\sigma = \sum_{\nu=1}^{n} m_\nu \eta_\nu$$

and taking

$$\omega_\sigma(\mathbf{t}) = \prod_{\nu=1}^{n} \Phi_\nu(\mathbf{t})^{\eta_\nu} \in K(\mathbf{t})$$

we obtain

$$F_\sigma(\mathbf{t})^{e_\sigma} = b_\sigma N_{K/k}(\omega_\sigma(\mathbf{t})) .$$

Hence by (12)

$$f(\mathbf{t}) = a \prod_{\sigma=1}^{s} b_\sigma N_{K/k}(\omega_\sigma(\mathbf{t})) = a_o N_{K/k}\left(\prod_{\sigma=1}^{s} \omega_\sigma(\mathbf{t})\right) .$$

By Theorem 34 there exist arithmetic progressions $P_1, \cdots, P_r$ such that if $\mathbf{t}^* \in P_1 \times \cdots \times P_r$ then $f(\mathbf{t}^*) \neq 0$. Take $\omega$ and $\mathbf{t}^* \in P_1 \times \cdots \times P_r$ such that (4) holds. We have

$$0 \neq N_{K/k}\omega^* = f(\mathbf{t}^*) = a_o N_{K/k}\left(\prod_{\sigma=1}^{s} \omega_\sigma(\mathbf{t}^*)\right)$$

$$\text{so} \quad a_o = N_{K/k}\left(\omega^* \prod_{\sigma=1}^{s} \omega_\sigma(\mathbf{t}^*)^{-1}\right) = N_{K/\ell}(\omega_o)$$

and we can satisfy (5) by taking $\omega(\mathbf{t}) = \omega_o \prod_{\sigma=1}^{s} \omega_\sigma(\mathbf{t})$. The proof is complete.

Remark. For $K/k$ of degree $p^2$ theorem 40 holds provided the multiplicities of factors are $\not\equiv 0 \pmod{p}$. (see Schinzel 1973a, for $k = \mathbb{Q}$). The following example shows that no assumption of this sort helps in general if $[K : k]$ has two distinct prime factors.

Example, $k = \mathbb{Q}$, $K = \mathbb{Q}\left(\sqrt{2\cos\frac{2\pi}{7}}\right)$. Let

$$f(t) = \left(t - 2\cos\frac{2\pi}{7}\right)\left(t - 2\cos\frac{4\pi}{7}\right)\left(t - 2\cos\frac{8\pi}{7}\right) =$$

$$= t^3 + t^2 - 2t - 1 .$$

The polynomial $f$ clearly is not a norm of an element of $K(t)$, since $[K : \mathbb{Q}] = 6$. Clearly $\sqrt{2\cos\frac{2\pi}{7}}$ is a zero of $f(t^2)$. We have (see Schinzel 1966)

$$\operatorname{disc} K = \operatorname{disc} f(t^2) = 2^6 \cdot 7^4$$

and the class number of K is one.

By Dedekind's theorem, any prime p factors in K in the same way as $f(t^2)$ factors mod p . Let $t^* \in \mathbb{Z}$ , then $f(t^*) = \pm \prod_{i=1}^{h} p_i^{\alpha_i}$ and since f(t) has a factor $t - t^*$ mod $p_i$ , $p_i$ has a factor of degree 1 in $L = \mathbb{Q}(2\cos\frac{2\pi}{7})$ . Let this factor be $\mathfrak{p}_i$ . One of the zeros of f is a quadratic residue mod $\mathfrak{p}_i$ since their product is 1 . Hence $f(t^2) \equiv 0 (\mathrm{mod}\, \mathfrak{p}_i)$ is solvable in L and since $\mathfrak{p}_i$ is of degree 1 in L the congruence $f(t^2) \equiv 0 (\mathrm{mod}\, p_i)$ is solvable in $\mathbb{Z}$ . Thus $p_i$ has a prime ideal factor of degree 1 in K , necessarily principal, say $(\pi_i)$ . We have

$$p_i = |N_{K/\mathbb{Q}}(\pi_i)| \; ; \quad \pm p_i = N_{K/\mathbb{Q}}(\pi_i) \; .$$

But fortunately

$$-1 = N_{K/k}\left(\sqrt{2\cos\frac{2\pi}{7}}\right) ,$$

hence with the proper choice of $\alpha_o$

$$f(t^*) = N_{K/\mathbb{Q}}\left(\sqrt{2\cos\frac{2\pi}{7}}^{\,\alpha_o} \prod_{i=1}^{h} \pi_i^{\alpha_i}\right) .$$

## Section 27. Norms of polynomials over a fixed field

**Definition 19.** *For a finite extension $K/k$, where $k$ is an algebraic number field, $P(K/k)$ denotes the set of prime ideals of $k$ that have a prime ideal factor of (relative) degree $1$ in $K$.*

**Theorem 41.** (Bauer 1916). *Let $k$ be an algebraic number field and $K/k$, $L/k$ two finite extensions. If $K/k$ is normal and $P(K/k) \supset P(L/k)$, then $L \supset K$. The same holds if the difference $P(L/k) \setminus P(K/k)$ is of (Dirichlet) density $0$.*

Proof. Let $K = k(\alpha)$, $\alpha$ an algebraic integer, $A$ the minimal polynomial of $\alpha$ over $k$. Take any prime ideal $\mathfrak{Q}$ of degree $1$ in $L$. If $\mathfrak{q}$ is the prime ideal of $k$ divisible by $\mathfrak{Q}$ we have $\mathfrak{q} \in P(L/k)$ so $\mathfrak{q} \in P(K/k)$ or $\mathfrak{q} \in S$, where $S = P(L/k) \setminus P(K/k)$ is of density $0$. Since $K/k$ is normal, if $\mathfrak{q} \in P(K/k)$ the polynomial $A(x)$ factors completely mod $\mathfrak{q}$ by Dedekind's theorem. Thus for all prime ideals $\mathfrak{Q}$ of degree 1 in $L$ except a set of density $0$ (consisting of divisors of the elements of $S$) $A(x)$ factors completely mod $\mathfrak{Q}$. By the Frobenius density theorem the Galois group of $A$ over $L$ consists only of the identity permutation. Hence $A(x)$ factors completely in $L$, whence, $\alpha \in L$ and $K \subset L$.

**Theorem 42.** (Davenport, Lewis and Schinzel 1964 for $k = \mathbb{Q}$, $r = 1$, Schinzel 1973a for $k = \mathbb{Q}$, $r > 1$). *Let $k$ be an algebraic number field, $K/k$ a normal extension and $F \in k[t_1, \cdots, t_r]$ a polynomial such that*

*(i) either $K/k$ is cyclic or the multiplicity of every irreducible factor of $F$ in $k$ is either prime to $[K : k]$ or divisible by it,*

*(ii) for every $r$ arithmetic progressions $P_1,\cdots,P_r$ in* **Z** *there exist integers $t_1^*,\cdots,t_r^*$ with $t_l^* \in P_l$ and an $\omega \in K$ such that*

$$(1) \qquad F(t_1^*,\cdots,t_r^*) = N_{K/k}(\omega) .$$

*Then*

$$(2) \qquad K(t_1,\cdots,t_r) = N_{K/k}(\Omega(t_1,\cdots,t_r)) \text{ for an } \Omega \in K[t_1,\cdots,t_r] .$$

<u>Lemma 1</u>. Let $K/k$ be normal, $B \in k[t]$ irreducible over $k$ with coefficients integral in $k$ and $e$ an integer prime to $d = [K : k]$. If for almost all prime ideals $\mathfrak{q}$ of $k$ such that $B(t) \equiv 0 \bmod \mathfrak{q}$ is solvable in **Z** there is an $\omega \in K$ satisfying

$$\operatorname{ord}_{\mathfrak{q}} N_{K/k}(\omega) = e$$

then

$$B(t) = b_1 N_{K/k}(B_1(t)) , \text{ for some } b_1 \in k , B_1 \in K[t] .$$

<u>Proof</u>. Let $B(\beta) = 0$ and let $L = k(\beta)$. We will prove that $S = P(L/k)\backslash P(K/k)$ has density $0$. Take $\mathfrak{q} \in P(L/k)$, $\mathfrak{q}$ of absolute degree 1 and not dividing the leading coefficient of $B$.

By Dedekind's theorem there exists an $x_0 \in k$ such that $B(x_0) \equiv 0 \bmod \mathfrak{q}$. Since $\mathfrak{q}$ is of absolute degree 1 the congruences $t \equiv x_0$ and $B(t) \equiv 0 \bmod \mathfrak{q}$ are solvable in **Z**. We have by the assumption

$$\operatorname{ord}_{\mathfrak{q}} N_{K/k}(\omega_0) = e .$$

Since $K/k$ is normal and $(d,e) = 1$ it follows that $\mathfrak{q}$ factors completely in $K$, where $\mathfrak{q} \in P(K/k)$. The set of prime ideals of $k$ of absolute degree greater than 1 has density 0, hence $P(L/k)\backslash P(K/k)$ also has density 0. By Theorem 4 we get $K \subset L$. If $K = k(\alpha)$, and $A$ is the minimal polynomial of $\alpha$ over $k$ then $A$ is a normal polynomial over $k$ and so factors completely in $L$. By Theorem 39 $B$ factors into $d$ factors in $K$. Hence $B \overset{\text{can}}{\underset{K}{=}} bB_1B_2\cdots B_d$ so $B = b_1N_{K/k}(B_1)$.

<u>Lemma 2</u>. Let $K/k$ be cyclic, $B \in k[t]$ irreducible with coefficients integral in $k$ and $e$ an integer. If for almost all prime ideals $\mathfrak{q}$ of $k$ for which $B(t) \equiv 0 \bmod \mathfrak{q}$ is solvable in $\mathbf{Z}$ there is an $\omega \in k$ satisfying $\mathrm{ord}_{\mathfrak{q}} N_{K/k}(\omega) = e$ then $B(t)^e = b_0 N_{K/k}(B_0(t))$ for some $B_0 \in K[t]$ and $b_0 \in k$.

Proof. Let $\mathfrak{q}$ be a prime ideal of $k$ unramified in $K$, such that the congruence $B(t) \equiv 0 \bmod \mathfrak{q}$ is solvable in $\mathbf{Z}$. By the assumption, apart from finitely many exceptional $\mathfrak{q}$'s there is an $\omega \in K$ such that

$$e = \mathrm{ord}_{\mathfrak{q}} N_{K/k}(\omega) .$$

Let $\mathfrak{q} = \mathfrak{Q}_1\mathfrak{Q}_2\cdots\mathfrak{Q}_g$ be the factorization of $\mathfrak{q}$ in $K$ and let $N_{K/k}\mathfrak{Q}_i = \mathfrak{q}^f$.

We have $f|e$, $f|[K:k] = d$, hence $f|(e,d)$. The field $K$ being cyclic over $k$ has a unique subfield $L \supset k$ of index $(e,d)$. $L$ is contained in the decomposition field of each $\mathfrak{Q}_i$, hence in $L$ the ideal $\mathfrak{q}$ factors into prime ideals of degree 1 over $k$. If $\lambda$ is an element of $L$ divisible by exactly one of those ideals (counting multiplicity), then

$$\operatorname{ord}_{\mathfrak{q}} N_{L/k}(\lambda^{e/(e,d)}) = \frac{e}{(e,d)} .$$

Now we can apply Lemma 1 since $[L:k] = \frac{d}{(e,d)}$ : this gives $B(t) = b_1 N_{L/k}(B_1(t))$ for some $b_1 \in k$, $B_1 \in L[t]$ and we can take $b_o = b_1^e$, $B_o = B_1^{e/(e,d)}$ .

Proof of Theorem 42. If $F(\mathbf{t}) = 0$ the theorem is trivial. If $F(\mathbf{t}) \neq 0$ let

$$(2) \qquad F(\mathbf{t}) \underset{k}{\overset{\text{can}}{=}} a \prod_{\sigma=1}^{s} F_\sigma(\mathbf{t})^{e_\sigma} ,$$

where $F_\sigma(\mathbf{t})$ have coefficients integral in $k$ and for a fixed $\sigma \leq s$ let $e_\sigma \not\equiv 0 \bmod [K:k]$ ,

$$(3) \qquad F_\sigma(\mathbf{t}) \underset{K}{\overset{\text{can}}{=}} c \prod_{\nu=1}^{n} \Phi_\nu(\mathbf{t}) ,$$

$$(4) \qquad N_{K/k}\Phi_\nu(t) = m_\nu F_\sigma(t)^{m_\nu} .$$

Since $\sigma$ is fixed we have omitted it is a subscript of $\Phi_\nu$ and $m_\nu$ . By Lemma 1 to Theorem 40 there exists an integral vector $\boldsymbol{\tau} \in Z^{r-1}$ such that the polynomials $\Phi_\nu(\boldsymbol{\tau},t)$ are irreducible in $K$ $(1 \leq \nu \leq n)$ and $B(t) = F_\sigma(\boldsymbol{\tau},t)$ , $e = e_\sigma$ satisfy the assumptions of Lemma 2 if $K/k$ is cyclic and of Lemma 1 otherwise. By these lemmata

$$(5) \qquad B(t)^e = b_o N_{K/k}(B_o(t)) ,\ b_o \in k ,\ B_o \in K[t] .$$

By (3) we have

$$B(t) \underset{K}{\overset{\text{can}}{=}} c \prod_{\nu=1}^{n} \Phi_\nu(\boldsymbol{\tau},t) .$$

Hence

$$B_o(t) = \beta \prod_{\nu=1}^{n} \Phi_\nu(\boldsymbol{\tau}, t)^{\eta_\nu}, \quad \eta_\nu \geq 0 (1 \leq \nu \leq n)$$

and by (4) and (5)

$$B(t)^{e_\sigma} = b_o N_{K/k} \beta \prod_{\nu=1}^{n} c_\nu^{\eta_\nu} B(t)^{m_\nu \eta_\nu} .$$

Thus

$$e_\sigma = \sum_{\nu=1}^{n} m_\nu \eta_\nu$$

and taking

$$\Omega_\sigma(\mathbf{t}) = \prod_{\nu=1}^{n} \Phi_\nu(\mathbf{t})^{\eta_\nu}, \quad b_\sigma = c^{e_\sigma} \prod_{\nu=1}^{n} c_\nu^{-\eta_\nu}$$

we obtain

$$F_\sigma(\mathbf{t})^{e_\sigma} = b_\sigma N_{K/k}(\Omega_\sigma(\mathbf{t})) .$$

The same is true with $b_\sigma = 1$, $\Omega_\sigma = F_\sigma^{e_\sigma/[K:k]}$ if $e_\sigma \equiv 0 \bmod [K:k]$. Hence by (2)

$$f(\mathbf{t}) = a \prod_{\sigma=1}^{s} b_\sigma N_{K/k}(\Omega_\sigma(\mathbf{t})) = a_o N_{K/k}(\prod_{\sigma=1}^{s} \Omega_\sigma(\mathbf{t})) .$$

By Theorem 34 there exist arithmetic progressions $P_1, \cdots, P_r$ such that if $\mathbf{t}^* \in P_1 \times \cdots \times P_r$ then $F(\mathbf{t}^*) \neq 0$. Take any $\mathbf{t}^* \in P_1 \times \cdots \times P_r$ such that $F(\mathbf{t}^*) = N_{K/k}(\omega^*)$. We have

$$0 \neq F(\mathbf{t}^*) = a_o N_{K/k}(\prod_{\sigma=1}^{s} \Omega_\sigma(\mathbf{t}^*)) ,$$

so

$$a_o = N_{K/k}(\omega^* \prod_{\sigma=1}^{s} \Omega_\sigma(\mathbf{t}^*)^{-1}) = N_{K/k}(\omega_o) .$$

and we can satisfy (1) by taking

$$\Omega(\mathbf{t}) = \omega_o \prod_{\sigma=1}^{s} \Omega_\sigma(\mathbf{t}) .$$

The proof is complete.

The following example shows that the assumption (i) in Theorem 42 is essential.

Example. $F(t) = t^2$, $k = \mathbb{Q}$, $K = \mathbb{Q}(\zeta_8)$. Every value of $F$ is a norm from $K/\mathbb{Q}$. Indeed, $F(0) = N_{K/\mathbb{Q}}(0)$, therefore assume $t^* \neq 0$ and let

$$t = \pm 2^\alpha \prod_{i=1}^{i_1} p_i^{\alpha_i} \prod_{i=1}^{i_2} q_i^{\beta_i} \prod_{i=1}^{i_3} r_i^{\gamma_i} ,$$

where $p_i, q_i, r_i$ are distinct primes satisfying the congruences $p_i \equiv 1 (\mathrm{mod}\ 4)$, $q_i \equiv 3\ (\mathrm{mod}\ 8)$, $r_i \equiv 7\ (\mathrm{mod}\ 8)$. Taking $K_1 = \mathbb{Q}(\zeta_4)$, $K_2 = \mathbb{Q}(\sqrt{-2})$, $K_3 = \mathbb{Q}(\sqrt{2})$ we get (see Mordell 1969, pp. 166 and 168):

$$2 = N_{K_1/\mathbb{Q}}(1 + \zeta_4) ,$$

$$p_i = a^2 + b^2 = N_{K_1/\mathbb{Q}}(a_i + b_i\zeta_4) ,$$

$$q_i = c_i^2 + 2d_i^2 = N_{K_2/\mathbb{Q}}(c_i + d_i\sqrt{-2}) ,$$

$$r_i = e_i^2 - 2f_i^2 = N_{K_3/\mathbb{Q}}(e_i + f_i\sqrt{2}) .$$

Since $K$ contains $K_1, K_2, K_3$ as subfields of index 2 we have

$$2^2 = N_{K/\mathbb{Q}}(1 + \zeta_4) ,$$

$$p_i^2 = N_{K/\mathbb{Q}}(a_i + b_i\zeta_4) ,$$

$$q_i^{\,2} = N_{K/\mathbb{Q}}(c_i + d_i\sqrt{-2}) , \quad r_i^2 = N_{K/\mathbb{Q}}(e_i + f_i\sqrt{2}) ;$$

whence

$$F(t^*) =$$

$$= N_{K/Q}((1+\zeta_4)^{\alpha} \prod_{i=1}^{i_1} (a_i + b_i\zeta_4)^{\alpha_i} \prod_{i=1}^{i_2} (c_i + d_i\sqrt{-2})^{\beta_i} \prod_{i=1}^{i_3} (e_i + f_i\sqrt{2})^{\gamma_i}).$$

But $F(t)$ is not a norm from $K[t]$ to $\mathbb{Q}(t)$ since its degree is not divisible by 4.

Corollary. Let $F \in \mathbb{Z}[t_1,\cdots,t_r]$ . If for every $r$ arithmetic progressions $P_1,\cdots,P_r$ there exist integers $t_1^*,\cdots,t_r^*,u,v$ , such that $t_\ell^* \in P_\ell$ and $F(t_1^*,\cdots,t_r^*) = u^2 + v^2$ then

$$F(t_1,\cdots,t_r) = U(t_1,\cdots,t_r)^2 + V(t_1,\cdots,t_r)^2$$

where $U , V \in \mathbb{Z}[t_1,\cdots,t_r]$ .

Proof. Let $\langle t_1,\cdots,t_r \rangle = \mathbf{t}$ . By Theorem 42 with $K = \mathbb{Q}(\zeta_4)$ we have $F(\mathbf{t}) = N(\Omega(\mathbf{t}))$ , where $\Omega \in \mathbb{Q}(\zeta_4)[\mathbf{t}]$ and $N$ denotes the norm from $\mathbb{Q}(\zeta_4,\mathbf{t})$ to $\mathbb{Q}(\mathbf{t})$ . Let $\Omega(\mathbf{t}) = c\,\Omega_o(\mathbf{t})$ , where $\Omega_o(\mathbf{t})$ is primitive. We get $F(\mathbf{t}) = N(\Omega(\mathbf{t})) = Nc\cdot N\,\Omega_o(\mathbf{t})$ , $N\,\Omega_o(\mathbf{t})$ has content 1, hence $Nc$ is the content of $F$ , denoted $C$ . Since $C \in \mathbb{Z}$ there exists an integer $c_o$ of $\mathbb{Q}(\zeta_4)$ such that

$$Nc_o = C .$$

Hence $F(\mathbf{t}) = N(c_o\Omega_o(\mathbf{t})) = U^2(\mathbf{t}) + V^2(\mathbf{t})$ , where

$U, V \in \mathbb{Z}[t]$ .

Remark. The corollary becomes false if the form $u^2 + v^2$ is replaced by the form $u^3 + v^3$ even if we assume that the latter represents $F(t^*)$ for all integers $t^* (r = 1)$ . A counterexample is given in Schinzel 1980. There remains the following

Problem. Suppose that $F(t) \in \mathbb{Z}[t]$ and for every $t^* \in \mathbb{Z}$ the equation $u^3 + v^3 = F(t^*)$ is solvable in $\mathbb{Z}$ . Do there exist polynomials $U, V \in \mathbb{Q}[t]$ that take integral values for every integer $t$ and satisfy
$U^3(t) + V^3(t) = F(t)$ ?

Even without the condition that $U, V$ be integer-valued the answer is not known to be affirmative, but the paper just quoted contains a conditional result in this direction.

## Section 28. Bauerian fields

**Definition 20.** *Let $k$ be an algebraic number field. A finite extension $K/k$ is Bauerian if for every finite extension $L/k$ the set $P(L/k)\setminus P(K/k)$ is of density $0$ only if $L \supset K'$ with $K'$ conjugate to $K'$ over $k$.*

**Corollary.** Every normal extension is Bauerian (by Theorem 41).

**Theorem 43.** (Schinzel 1966 for $k = \mathbb{Q}$). *An extension $K/k$ is Bauerian if and only if the group $\mathcal{G}$ of the normal closure of $K$ over $k$ represented as a permutation group on the conjugates of $K$ over $k$ has the following property:*

*Any subgroup $\mathcal{H}$ of $\mathcal{G}$ contained in the union of stability subgroups is contained in one of them.*

Proof. Let $K = k(\alpha)$, $\alpha$ be an algebraic integer and $A(x)$ the minimal polynomial of $\alpha$ over $k$. Then $\mathcal{G}$ is the group of $A(x)$ over $k$.

Necessity. Take a subgroup $\mathcal{H}$ contained in the union of stability subgroups. Take the field $L$ invariant under $\mathcal{H}$. Then $\mathcal{H}$ is the group of $A(x)$ over $L$. Take any $\mathfrak{q} \in P(L/k)$ not dividing the discriminant of $A$. It has a prime ideal factor $\mathfrak{Q}$ of degree $1$ in $L$. Every element of $\mathcal{H}$ fixes some zero of $A$. Thus (see Mann 1955, Theorem 16.6) $A(x) \equiv 0 \bmod \mathfrak{Q}$ is solvable in $k$ and so is $A(x) \equiv 0 \bmod \mathfrak{q}$, whence $\mathfrak{q} \in P(K/k)$. Therefore $P(L/k)\setminus P(K/k)$ is finite, whence $L \supset K'$, hence $A(x)$ has a zero in $L$ and $\mathcal{H}$ is in one of the stability subgroups.

Sufficiency. Take any $L$ and consider the subgroup $\mathcal{H}$

of $\mathcal{G}$ corresponging to the intersection $\tilde{K} \cap L$ , where $\tilde{K}$ is the normal closure of $K$ over $k$ . Clearly $\mathcal{J}$ is the group of $A(x)$ over $L$ . Suppose that an element of $\mathcal{J}$ fixes no zero of $A$ . Then by the Frobenius density theorem there is a set $T$ of prime ideals of degree $1$ of $L$ with positive density and with the property that $A(x) \equiv 0 \bmod \mathfrak{Q}$ is unsolvable in $k$ for $\mathfrak{Q} \in T$ . If $\mathfrak{Q} \in T$ and $\mathfrak{q}$ is the prime ideal of $k$ divisible by $\mathfrak{Q}$ then $A(x) \equiv 0 \bmod \mathfrak{q}$ is unsolvable in $k$ , hence $\mathfrak{q} \notin P(K/k)$ . Thus if $P(L/k)\setminus P(K/k)$ is of density $0$ then every element of $\mathcal{J}$ leaves some element fixed. Therefore $\mathcal{J}$ is in a stability subgroup, $A(x)$ has a zero in $L$ and $L \supset K'$ with $K'$ conjugate to $K$ over $k$ . The proof is complete.

<u>Theorem 44</u> (Schinzel 1973a for $k = \mathbb{Q}$). *A Bauerian extension $K/k$ can be substituted for a normal extension in Theorem 42 if and only if the Galois group $\mathcal{G}$ of the normal closure of $K$ over $k$ represented as a permutation group on the conjugates of $K$ over $k$ has the following property: every permutation of $\mathcal{G}$ that decomposes into cycles of coprime lengths fixes at least one field.*

<u>Proof</u>. <u>Necessity</u>. Take a permutation $\sigma$ with cycles of coprime lengths and let $\mathcal{J} = \langle \sigma \rangle$ , and $L$ be the field invariant under $\mathcal{J}$ . Let $K = k(\alpha)$ , $L = k(\beta)$ and let $A$ , $B$ be the minimal and hence monic polynomials over $k$ of $\alpha$ and $\beta$ respectively. It is clear that

$$A(t) \underset{L}{\overset{\text{can}}{=}} \prod_{\rho=1}^{r} A_\rho(t) ,$$

where the sets of the zeros of $A_\rho (\rho = 1,\cdots,r)$ are orbits of $<\sigma>$. Hence the lengths of cycles of $\sigma$ are $|A_1|,\cdots,|A_r|$.

By Theorem 39 we have

$$(1) \qquad B(t) \underset{K}{\overset{\text{can}}{\equiv}} \prod_{\rho=1}^{r} B_\rho(t) ,$$

where

$$\frac{|B_\rho|}{|B|} = \frac{|A_\rho|}{|A|} = \frac{|A_\rho|}{[K:k]}$$

and where $B_\rho$ are monic. By Lemma 2 to Theorem 33 we get

$$(2) \qquad N_{K/k} B_\rho(t) = B(t)^{|A_\rho|} .$$

Since $(|A_1|,\cdots,|A_r|) = 1$ we can find integers $a_1,\cdots,a_r$ such that

$$\sum_{\rho=1}^{r} a_\rho |A_\rho| = 1 .$$

Then

$$B(t) = N_{K/k} (\prod_{\rho=1}^{r} A_\rho(t)^{a_\rho})$$

and for every integer $t^*$, $B(t^*)$ is a norm from $K$. If Theorem 42 is to hold $B(t)$ must be a norm from $K[t]$ thus by (1) and (2) $|A_\rho| = 1$ for some $\rho \leq r$ and $\sigma$

fixes at least one field.

Sufficiency. It is enough to show that Lemma 1 to Theorem 42 holds. Let $K = k(\alpha)$, $L = k(\beta)$, $A$, $B$ be the minimal polynomials over $k$ of $\alpha, \beta$ respectively, $d = [K:k]$. If $(e,d) = 1$ and for a prime ideal $\mathfrak{q}$ of $k$ there is an $\omega \in K$ satisfying $\mathrm{ord}_{\mathfrak{q}} N_{K/k}(\omega) = e$ then $\mathfrak{q}$ factors in $K$ into prime ideals of relatively prime degrees $f_1, \cdots, f_r$.

Let $\mathfrak{Q}$ be a factor of $\mathfrak{q}$ in the normal closure $\tilde{K}$ of $K$ over $k$. The Frobenius substitution $\sigma$ satisfying $\gamma^{\sigma} \equiv \gamma^{N\mathfrak{q}} \bmod \mathfrak{Q}$ for all integers $\gamma$ of $\tilde{K}$ factors into cycles of lengths $f_1, \cdots, f_r$ (see Mann 1955, Theorem 16.6). Hence by our condition one of the numbers $f_i$ is 1. Thus $\mathfrak{q} \in P(K/k)$. If this happens for almost all prime ideals $\mathfrak{q}$ of $k$ for which $B(t) \equiv 0 \bmod \mathfrak{q}$ is solvable in $Z$ we infer that $P(L/k) \setminus P(K/k)$ is of density 0. Since $K/k$ is Bauerian this implies $L \supset K'$ for a conjugate $K'$ of $K$ over $k$. Thus $A(t)$ has in $L$ a linear factor and by Theorem 39 $B(t)$ has in $K$ a factor $B_1$ of degree $\frac{|B|}{|A|}$. By Lemma 2 to Theorem 33 we get

$$b_1 N_{K/k} B_1 = B .$$

Remark. Examples of Bauerian fields satisfying the conditions of Theorem 44 are furnished by all cubic and quartic extensions and all primitive solvable extensions of degree $p^n$, where $p$ is a prime and $n \leq 5$ (see Schinzel 1973a).

It has been shown by R. Griess and R. Lyons (unpublished) that not all primitive solvable extensions are

Bauerian; a counterexample is an extension of degree 64 with the Galois group $SU(3,\mathbb{F}_2)$. On the other hand there are Bauerian extensions that do not satisfy the conditions of Theorem 44, for an example see Schinzel 1973a.

# References

M.Artin and D.Mumford 1972, Some elementary examples of unirational varieties which are non-rational, Proc. London Math.Soc. (3) 25, pp. 75-95.

M.Bauer 1916, Zur Theorie der algebraischen Zahlkörper, Math. Ann. 77, pp. 353-56.

M.Bauer 1929, Bemerkung zur Algebra, Acta Litt. Scient. (Szeged), Sect. Sc. Math. 4, pp. 244-245.

A.Bazylewicz 1976, On the product of the conjugates outside the unit circle of an algebraic integer, Acta Arith. 30, pp. 43-61.

E.Bertini 1882, Sui sistemi lineari, Rend. Ist. Lombardo (2) 15, pp. 24-28.

G.A.Bliss 1933, Algebraic functions, Amer. Math. Soc. Colloq. Publ. 16, New York.

A.Bremner and P.Morton 1978, Polynomial relations in characteristic p, Quarterly J.Math. Oxford. Ser. (2) 29, pp. 336-347.

A.Capelli 1897, Sulla riduttibilità delle equazioni algebriche. Nota prima, Rend. Accad. sc. fis. mat, Soc. Napoli (3), 3, pp. 243-252.

A.Capelli 1898, Sulla riduttibilità delle equazioni algebriche. Nota secunda, ibidem, (3) 4, pp. 84-90.

G.Castelnuovo 1894, Sulla rationalità della involuzioni piane. Math. Ann. 44, pp. 125-155.

C.H.Clemens, Jr and P.A.Griffiths 1972, The intermediate Jacobian of the cubic threefold, Ann. Math. 95, pp. 281-356.

H.Davenport, D.J.Lewis and A.Schinzel 1964, Polynomials of certain special types, Acta Arith. 9, pp. 107-116, =H.Davenport, Collected papers, vol. 4. Academic Press, London-New York, San Francisco 1977, pp. 1720-1729.

H.Davenport, D.J.Lewis and A.Schinzel 1966, Quadratic diophantine equations with a parameter, Acta Arith. 11, pp. 353-358, =H.Davenport, Collected papers, vol. 4, Academic Press, London-New York-San Francisco 1977, pp. 1730-1735.

L.E.Dickson 1926, Modern algebraic theories, Sanborn and Co., Chicago, New York, etc.

E.Dobrowolski 1979, On a question of Lehmer and the number of irreducible factors of a polynomial, Acta Arith. 34, pp. 391-401.

F.Dorey and G.Whaples 1974, Prime and composite polynomials, J.Algebra 28, pp. 88-101.

K.Dörge 1972, Einfacher Beweis des Hilbertschen Irreduzibilitätssatzes, Math. Ann. 96, pp. 176-182.

M.Eichler 1939, Zum Hilbertschen Irreduzibilitätsatz. Math. Ann. 116, pp. 742-748.

H.T.Engstrom 1941, Polynomial substitutions, Amer. J. Math. 63, pp. 249-255.

E.Fischer 1925, Uber absolute Irreduzibilität (Aus einem Briefe an E.Noether), Math. Ann. 94, pp. 163-165.

W.Franz 1931, Untersuchungen zum Hilbertschen Irreduzibilitätsatz, Math. Z. 33, pp. 275-293.

M.Fried 1970, On a conjecture of Schur, Michigan Math. J. 17, pp. 41-50.

M.Fried 1973, The field of definition of function fields and a problem in the reducibility of polynomials in two variables, Illinois J. Math. 17, pp. 128-146.

M.Fried 1974a, On a theorem of Ritt and related diophantine problems. J. Reine Angew. Math. 264, pp. 40-55.

M.Fried 1974b, On Hilbert's irreducibility theorem, J. Number Theory 6, pp. 211-232.

M.Fried and R.E.Mac Rae 1969, On the invariance of chains of fields, Illinois J. Math. 13, pp. 165-171.

M.Fried and A.Schinzel 1972, Reducibility of quadrinomials, Acta Arith. 21, pp. 153-171.

J.Vicente Goncalvez 1956, L'inégalité de W. Specht, Univ. Lisboa Revista Fac. Ci (2) A 1, pp. 167-171.

P.Gordan 1887, Über biquadratische Gleichungen, Math. Ann. 29, pp. 318-326.

E.Gourin 1930, On Irreducible Polynomials in Several Variables which Become Reducible when the Variables are Replaced by Powers of Themselves, Trans. Amer. Math. Soc. 32, pp. 485-501.

H.Hasse 1930, Bericht über neuere Untersuchungen und Probleme aus der Theorie der algebraischen Zahlkörper. Teil II Reziprozitätsgesetz, 2 Auflage, Physica-Verlag, Würzburg-Wien 1965.

H.Hasse 1932, Zwei Bemerkungen zu der Arbrit "Zur Arithmetik der Polynome" von U.Wagner in den Mathematischen Annalen Bd 105, S. 628-631, Math. Ann. 106, pp. 455-456.

D.Hilbert 1892, Über die Irreduzibilität ganzer rationalen Functionen mit ganzzahligen Coeffizienten, J.Reine Angew. Math. 110, pp. 104-129 = Ges. Abhandlungen Bd II, Springer 1970, pp. 264-286.

A.Hurwitz 1894, Über die Theorie der Ideale. Nachr. K. Ges. Wiss. Göttingen, pp. 291-298 =Math. Werke, Bd II, Birkhäuser, Basel 1933, pp. 191-197.

A.Hurwitz 1895, Über einen Fundamentalsatz der arithmetischen Theorie der algebraischen Grössen, Nachr. K. Ges. Wiss. Göttingen, pp. 230-240, =Math.Werke, Bd II, Birkhäuser, Basel 1933, pp. 198-207.

A.Hurwitz 1913, Über die Trägheitsformen eines algebraischen Moduls, Ann. Mat. Pura Appl. (3) 20 pp. 113-151 =Math. Werke, Bd II, Birkhäuser, Basel 1933, pp. 591-626.

J.Igusa 1951, On a theorem of Lueroth, Mem. Coll. Sci. Univ. Kyoto, Ser. A, Math. 26, pp. 251-253.

V.A.Iskovskih and Yu.I.Manin 1971, Threedimensional quartics and counterexamples to the Lüroth problem, Math. Sb. 86, pp. 140-166. (Russian).

H.Kapferer 1929, Über Resultanten und Resultanten-Systeme, Sitzungsber, Bayer. Akad. München, pp. 179-200.

A.Kneser 1887, Über die Gattung niedrigster Ordnung unter welcher gegebene Gattungen algebraisches Grössen enthalten sind, Math. Ann. 30, pp. 196-202.

M.Kneser 1975, Lineare Abhängigkeit von Wurzeln, Acta Arith. 26, pp. 307-308.

T.Kojima 1915, Note on Number-theoretic Properties of Algebraic Functions, Tohôku Math. J. 8, pp. 24-37.

J.König 1903, Einleitung in die allgemeine Theorie der algebraischen Grössen, Teubner, Leipzig.

L.Kronecker 1882, Grudzüge einer arithmetischen Theorie der algebraischen Grössen, J. Reine Angew. Math. 92, pp. 1-122 = Werke **2, Chelsea 1968, pp. 237-387.**

L.Kronecker 1883, Zur Theorie der Formen höherer Stufen, Monatsber. Akad. Wiss. Berlin 37, pp. 957-960 =Werke **2, Chelsea 1968, pp. 417-424.**

W.Krull 1937, Über einen Irreduzibilitätssatz von Bertini, J. Reine Angew. Math. 177, pp. 94-104.

E.Landau 1905, Sur quelques théorèmes de M.Petrovich relatifs aux zéros des fonctions analytiques, Bull. Soc. Math. de France, 33, pp. 1-11.

S.Lang 1962, Diophantine geometry, Interscience Publishers, New York-London.

H.Levi 1942, Composite polynomials with coefficients in an arbitrary field of characteristic zero, Amer. J. Math. 64, pp. 389-400.

D.J.Lewis and A.Schinzel 1980, Quadratic diophantine equations with parameters, Acta Arith 37, pp. 133-141.

P.Lüroth 1876, Beweis eines Satzes über rationale Curven, Math. Ann. 9, pp. 163-165.

F.S.Macaulay 1916, The algebraic theory of modular systems, 2 nd edition, Stechert and Hafner, New York and London 1964.

H.B.Mann 1955, Introduction to algebraic number theory, The Ohio State University Press, Columbus

F.Mertens 1899, Zur Eliminationtheorie, Sitzungsber. K. Akad. Wiss. Wien, Math. Naturw. Kl. 108, pp. 1178-1228, pp.1244-1386.

F.Mertens 1911, Über die Zerfällung einer ganzen Funktion einer Veränderlichen in zwei Faktoren, ibid. 120, pp. 1485-1502.

T.T.Moh and W.J.Heinzer, A generalized Lüroth Theorem for Curves, Japanese J. Math. 31 (1979), pp. 85-86.

H.L.Montgomery and A.Schinzel 1977, Some arithmetic properties of polynomials in several variables, in the book: Transcendence theory: Advances and applications, Academic Press, London, New York, San Francisco, pp. 195-204.

L.J.Mordell 1953, On the linear independence of algebraic numbers, Pacific J. Math. 3, pp. 625-630.

L.J.Mordell 1969, Diophantine equations, Academic Press, London, New York, San Francisco.

E.Netto 1895, Über einen Lüroth-Gordanschen Satz, Math. Ann. 46, pp. 310-318.

E.Noether 1915, Körper und Systeme rationaler Funktionen, Math. Ann. 76, pp. 161-196.

E.Noether 1922, Ein algebraisches Kriterium für absolute Irreduzibilität, Math. Ann. 85, pp. 26-33.

A.Ostrowski 1919, Zur arithmetischen Theorie der algebraischen Grössen, Nachr. K. Ges. Wiss. Göttingen, pp. 273-298.

A.Ostrowski 1960, On an inequality of J. Vincente Goncalvez, Univ. Lisboa Revista Fac. Ci (2) A 8, pp.115-119.

O.Perron 1951, Algebra vol. 1, 3 rd ed, de Gruyter, Berlin.

G.Polya and G.Szegö 1925, Aufgaben und Lehrsätze aus der Analysis, Bd II, Julius Springer, Berlin.

H.Prüfer 1932, Untersuchungen über Teibarkeitseigenschaften in Körpern, J. Reine Angew. Math. 168, pp. 1-36.

V.Puiseux 1850, Recherches sur les fonctions algébriques, J. Math. Pure Appl. 15, pp. 365-480.

A.Rédei 1959, Algebra, I Teil, Akademische Verlaggesellschaft, Leipzig.

A.Riehle 1919, Über den Bertinischen Satz und seine Erweiterung, Diss. Tübingen.

J.F.Ritt 1922, Prime and Composite Polynomials, Trans. Amer. Math. Soc. 23, pp. 51-66.

J.F.Ritt 1927, A factorization theory for functions $\Sigma_{i=1}^{n} a_i e^{\alpha_i x}$, Trans. Amer. Math. Soc. 29, pp. 584-596.

G.Salomon 1915, Über das Zerfallen von Systemen von Polynomen, Jhber. Deutsche Math. Ver. 24, pp. 225-246.

P.Samuel 1953, Some remarks on Lüroth's Theorem, Mem. Coll. Sci. Univ. Kyoto, Ser. A. Math. 27, pp. 223-224.

A.Schinzel 1963a, Some unsolved problems on polynomials, Math. Bibliteka 25, pp. 63-70.

A.Schinzel 1963b, Reducibility of polynomials in several variables, Bull. Acad. Polon. Sci. Ser. sci. math. astr. phys. 11, pp. 633-638.

A.Schinzel 1965, On Hilbert's Irreducibility Theorem, Ann. Polon. Math. 16, pp. 333-340.

A.Schinzel 1966, On a theorem of Bauer and some of its applications, Acta Arith. 11, pp. 333-344., Corrigendum ibid. 12 (1967), p.425.

A.Schinzel 1973a, On a theorem of Bauer and some of its applications II, Acta Arith. 22, pp. 222-231.

A.Schinzel 1973b, A general irreducibility criterion, J. Indian Math. Soc. 37, pp. 1-8.

A.Schinzel 1975a, On the product of the conjugates outside the unit circle of an algebraic number, Addendum, Acta Arith. 26, pp. 329-331.

A.Schinzel 1975b, On linear dependence of roots, Acta Arith. 28, pp. 161-175.

A.Schinzel 1976, On the number of irreducible factors of a polynomial, Colloq. Math. Soc. János Bolyai 13, pp. 305-314.

A.Schinzel 1977, Abelian binomials, power residues and exponential congruences, Acta Arith. 32, pp. 245-274.

A.Schinzel 1978, Reducibility of lacunary polynomials III, Acta Arith. 34, pp. 227-266.

A.Schinzel 1980, On the relation between two conjectures on polynomials, Acta Arith. 38, pp. 285-322.

A.Schinzel 1981, **Families of curves having each an integer point, Acta Arith. 40.**

H.A.Schwarz 1882, Démonstration élémentaire d'une propriété fondamentale des fonctions interpolaires, Atti R. Accad. Sc. Torino 17, pp. 740-742 = Ges. Math. Abh., 2 Band, Berlin 1890, pp. 307-308.

B.Segre 1946, Sui sistemi continui di ipersuficie algebriche, Rend. Accad. Naz. Lincei (8) 1, pp. 564-570.

B.Segre 1951, Sull esistenza, sia nel campo rationale che nel campo reale, di involuzioni piane non--birazionali, Rend. Accad. Naz. Lincei (8) 10, pp.564-570.

C.L.Siegel 1929, Über einige Anwendungen diophantischer Approximationen, Abh. Preuss. Akad. Phys. Math. Klasse, pp. 41-69 = Ges. Abh. Bd I, Springer pp. 209-266.

C.L.Siegel 1972, Algebraische Abhängigkeit von Wurzeln, Acta Arith. 21, pp. 59-64.

T.Skolem 1940, Einige Sätze über Polynome, Avhandlinger Norske Vid. Akad. Oslo, I Mat-Naturv. Kl. No 4.

C.J.Smyth 1971, On the product of the conjugates outside the unit circle of an algebraic integer, Bull. London Math. Soc. 3, pp. 169-175.

W.Specht 1950, Abschätzungen der Wurzeln algebraischer Gleichungen, Math. Z. 52, pp. 316-321.

V.G.Sprindzuk 1979, Hilbert's irreducibility theorem and rational points on algebraic curves, Dokl. ANSSSR 247, pp. 285-289 (Russian).

E.Steinitz 1910, Algebraische Theorie der Körper, J. Reine Angew. Math. 137, pp. 167-309 = Abdrück, Berlin-Leipzig 1930.

H.Tverberg 1966, On the irreducibility of polynomials f(x) + g(y) + h(z), Quart. J. Math. Oxford Ser. (2) 17, pp. 364-366.

B.L.van der Waerden 1926, Ein algebraisches Kriterium für die Lösbarkeit eines Systems homogener Gleichungen, Proc. Kon. Akad. Wet. 29, pp. 142-149.

B.L.van der Waerden 1928, Neue Begründung der Eliminations-und Resultantentheorie, Nieuw Arch. Wisk. (2) 15, pp. 302-320.

B.L.van der Waerden 1937, Zur algebraischen Geometrie X. Über lineare Scharen von reduziblen Mannigfaltigkeiten, Math. Ann. 113, pp. 705-712.

B.L.van der Waerden 1970, Algebra, vol. 1, Frederick Ungar, New York.

O.Zariski 1941, Pencils on an algebraic variety and a new proof of a theorem of Bertini, Trans. Amer. Math. Soc. 50, pp. 48-70. =Selected Papers, vol. 1, MIT Press, Cambridge Mass. 1972, pp. 154-176.

O.Zariski 1958, On Castelnuovo's criterion of rationality $p_a = P_2 = 0$ of an algebraic surface, Illinois. J. Math. 2, pp. 305-315.

# Appendix

Theorem 3 (see page 8) also holds for a finite field $k$. The following proof has been found by F. Laubie and the author, see their note in C.R. Acad. Sci. France, 1981 entitled "Sur le théoreme de Gordan-Igusa".

Let $k$ be a finite field. For such a field, the proof of Theorem 3 given in Section 3 gives only the existence of a finite extension $k_o$ of $k$ such that $k_oK = k_o(g_o)$, where $g_o$ is in $k_o(x_1,\ldots,x_n)$. Let $g_o = P/Q$, where $P, Q \in k[x_1,\ldots,x_n]$, $(P,Q) = 1$. Since $g_o \notin k_o$, there exist monomials $M_1$ and $M_2$ such that the coefficients $p_i, q_i$ of $M_i$ in $P$ and $Q$ respectively satisfy $p_1q_2 - q_1p_2 \neq 0$. Now let $\sigma$ be the substitution which generates the Galois group of $k_1/k$. It operates in the obvious way on $K_o[x_1,\ldots,x_n]$ and we have $g_o = P^\sigma/Q^\sigma$. On the other hand $k_o(g_o^\sigma) = k_o^\sigma(g_o^\sigma) = (k_o(g_o))^\sigma = (k_oK)^\sigma = k_oK = k_o(g_o)$, hence

$$g_o^\sigma = \frac{ag_o + b}{cg_o + d} = \begin{pmatrix} a & b \\ c & d \end{pmatrix} g_o, \text{ where } a,b,c,d \in k_o \text{ and } \begin{vmatrix} a & b \\ c & d \end{vmatrix} \neq 0.$$

Since $\dfrac{aP + bQ}{cP + dQ} = \dfrac{P^\sigma}{Q^\sigma}$ and $(P^\sigma, Q^\sigma) = 1 = (aP + bQ, cP + dQ)$, we have for suitable $e \in k_o$ that $aP + bQ = eP^\sigma$, $cP + dQ = eQ^\sigma$. Comparing the coefficients of the monomial $M_i$ on both sides we obtain

$$ap_i + bq_i = ep_i^\sigma, \quad cp_i + dq_i = eq_i^\sigma$$

which gives

$$\begin{pmatrix} a & b \\ c & d \end{pmatrix} \begin{pmatrix} p_1 & p_2 \\ q_1 & q_2 \end{pmatrix} = \begin{pmatrix} p_1^\sigma & p_2^\sigma \\ q_1^\sigma & q_2^\sigma \end{pmatrix} \begin{pmatrix} c & o \\ o & c \end{pmatrix} .$$

Putting $g = \begin{pmatrix} p_1 & p_2 \\ q_1 & q_2 \end{pmatrix}^{-1} g_o$ we find

$$g^\sigma = \begin{pmatrix} p_1^\sigma & p_2^\sigma \\ q_1^\sigma & q_2^\sigma \end{pmatrix}^{-1} g_o^\sigma = \begin{pmatrix} p_1^\sigma & p_2^\sigma \\ q_1^\sigma & q_2^\sigma \end{pmatrix}^{-1} \begin{pmatrix} a & b \\ c & d \end{pmatrix} g_o$$

$$= \begin{pmatrix} p_1^\sigma & p_2^\sigma \\ q_1^\sigma & q_2^\sigma \end{pmatrix} \begin{pmatrix} a & b \\ c & d \end{pmatrix} \begin{pmatrix} p_1 & p_2 \\ q_1 & q_2 \end{pmatrix} q = \begin{pmatrix} e & o \\ o & e \end{pmatrix} q \; .$$

Hence $g \in K$ and since $k_o K = k_o(g_o)$ and $[k_o K : K] = [k_o : k]$ we get $K = k(g)$ .

Printed and bound by CPI Group (UK) Ltd, Croydon, CR0 4YY

16/04/2025

14658539-0002